CAMBRIDGE LIBRARY COLLECTION

Books of enduring scholarly value

Darwin

Two hundred years after his birth and 150 years after the publication of 'On the Origin of Species', Charles Darwin and his theories are still the focus of worldwide attention. This series offers not only works by Darwin, but also the writings of his mentors in Cambridge and elsewhere, and a survey of the impassioned scientific, philosophical and theological debates sparked by his 'dangerous idea'.

Essay on the Theory of the Earth

Essay on the Theory of the Earth was the last work of the scientific writer Robert Kerr who translated it from the introductory essay of George Cuvier's four-volume Recherches sur les ossements fossiles de quadrupèdes. Before its first publication in 1813, the essay was partly expanded by the geologist and natural historian Robert Jameson who wrote a preface and included extensive notes on mineralogy. Using geological evidence as its principal source of enquiry, Cuvier's essay attempts to address the questions of the origins of the human race, the formation of the earth, and the correlation between incomplete fossil remains and existing species of animals. Extremely influential in its own time, the essay remains a source of considerable insight into the early development of geological research, examining issues of continued significance today.

Essay on the Theory of the Earth

Georges Cuvier

CAMBRIDGE UNIVERSITY PRESS

Cambridge New York Melbourne Madrid Cape Town Singapore São Paolo Delhi

Published in the United States of America by Cambridge University Press, New York

www.cambridge.org
Information on this title: www.cambridge.org/9781108005555

This edition first published 1815
This digitally printed version 2009

ISBN 978-1-108-00555-5

ESSAY

ON THE

THEORY OF THE EARTH.

ESSAY

ON THE

THEORY OF THE EARTH.

ESSAY

ON THE

Theory of the Earth.

TRANSLATED FROM THE FRENCH OF

M. CUVIER,

PERPETUAL SECRETARY OF THE FRENCH INSTITUTE, PROFESSOR AND ADMINISTRATOR OF THE MUSEUM OF NATURAL HISTORY,

&c. &c.

BY

ROBERT KERR, F.R.S. & F.A.S. EDIN.

WITH

MINERALOGICAL NOTES,

AND

AN ACCOUNT OF CUVIER'S GEOLOGICAL DISCOVERIES,

BY PROFESSOR JAMESON.

SECOND EDITION, WITH ADDITIONS.

EDINBURGH:

PRINTED FOR WILLIAM BLACKWOOD, SOUTH BRIDGE STREET, EDINBURGH; AND JOHN MURRAY, ALBEMARLE-STREET, AND ROBERT BALDWIN, PATERNOSTER-ROW, LONDON.

1815.

PREFACE.

Although the Mosaic account of the creation of the world is an inspired writing, and consequently rests on evidence totally independent of human observation and experience, still it is interesting, and in many respects important, to know that it coincides with the various phenomena observable in the mineral kingdom. The structure of the earth, and the mode of distribution of extraneous fossils or petrifactions, are so many direct evidences of the truth of the scripture account of the formation of the earth; and they might be used as proofs of its author having been inspired, because the mineralogical facts discovered by modern naturalists were unknown to the sacred historian. Even the periods of time, the six days of the Mosaic description, are not inconsistent with our theories of the earth. There are, indeed, many physical considerations which render it probable that the motions of the earth may have been slower during the time of its formation than after it was formed,* and consequently that the day, or period between morning and evening, may have then been indefinitely longer than it is at present. If such a hypothesis is at all admissible, it will go far in supporting the opinion which has long been maintained on this subject by many of the ablest and most learned scripture critics. The deluge, one of the grandest natural events described in the Bible; is equally confirmed, with regard to its extent and the period of its occurrence, by a careful study of the various phenomena observed on and near the earth's surface. The age of the human

---

* *Vide* Bishop Horsley's Sermons, p. 445. Second edition.

race, also a most important enquiry, is satisfactorily determined by an appeal to natural appearances; and the pretended great antiquity of some nations, so much insisted on by certain philosophers, is thereby shewn to be entirely unfounded.

These enquiries, particularly what regards the *deluge*, form a principal object of the Essay of the illustrious German naturalist Cuvier, now presented to the English reader.* After describing the principal results at which the theory of the earth, in his opinion, has arrived, he next mentions the various relations which connect the history of the fossil bones of land animals with these results; explains the principles on which is founded the art of ascertaining these bones, or, in other words, of discovering a genus, and of distinguishing a species, by a single fragment of bone; and gives a rapid sketch of the results to which his researches lead, of the new genera and species which these have been the

* Cuvier is a native of Mumpelgardt in Germany.

means of discovering, and of the different formations in which they are contained. Some naturalists, as La Mark, having maintained that the present existing races of quadrupeds are mere modifications or varieties of those ancient races which we now find in a fossil state, modifications which may have been produced by change of climate, and other local circumstances, and since brought to the present great difference by the operation of similar causes during a long succession of ages,—Cuvier shews that the difference between the fossil species and those which now exist, is bounded by certain limits; that these limits are a great deal more extensive than those which now distinguish the varieties of the same species; and, consequently, that the extinct species of quadrupeds are not varieties of the presently existing species. This very interesting discussion naturally leads our author to state the proofs of the recent population of the world; of the comparatively modern origin of its present surface; of the deluge, and the subsequent renewal of human society. Subjects so important, and treated by one of the first philosophers of the age, a man not less distinguished for extent and accuracy of knowledge, than for originality of views, and elegance of style, cannot fail to excite very general notice, to fix the attention of the naturalist on a new series of facts, to admonish the sceptic, and afford the highest pleasure to those who delight in illustrating the truth of the Sacred Writings, by an appeal to the facts and reasonings of natural history.

This Translation was executed by a gentleman well know to the philosophical world by his various useful writings, the late Mr Kerr, whom a sudden death has snatched from this transitory scene. The few notes I have added, will, I trust, be found interesting; and the short account of Cuvier's Geological Discoveries, which accompanies them, will be useful to those who have not an opportunity of consulting the great work.

ROBERT JAMESON.

ADVERTISEMENT

TO

THE SECOND EDITION.

It was my intention to have prefixed to this edition of the Essay on the Theory of the Earth, a general discourse on Geognosy, with the view of explaining its various relations to the other branches of Natural History, and of examining some of the criticisms that have appeared on the work itself, and the notes and preface accompanying it, but want of leisure has prevented me at present from carrying into effect this plan. I have however enlarged the Notes, and added a series of observations on a very interesting topic; the distribution of fossil organic remains throughout the crust of the Earth, and also as intimately connected with the discussions contained in the Essay of Cuvier; an account of the Mineralogy of the country round Paris; and of some remarkable mineral formations that occur in the South of England.

ROBERT JAMESON.

CONTENTS.

MINERALOGICAL NOTES AND ILLUSTRATIONS.

PLATE I.

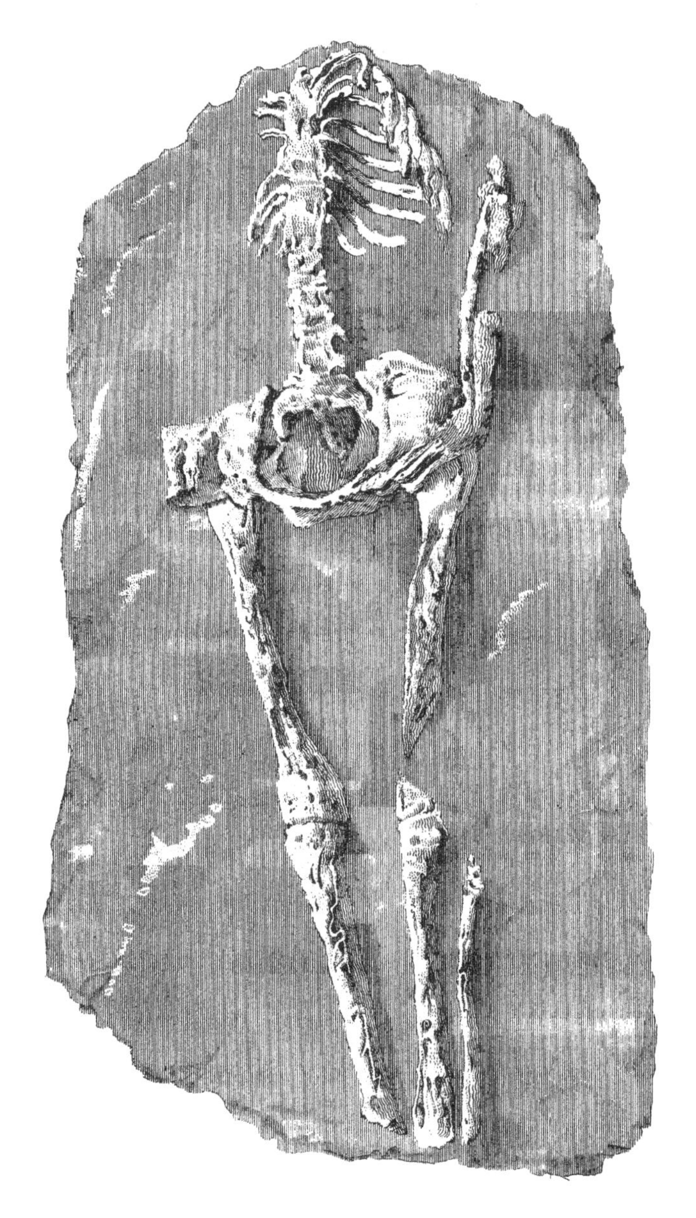

FOSSIL

HUMAN SKELETON.

FOUND IN GUADALOUPE

Engd by W. & D. Lizars Edinr

Edinburgh Published by Wm Blackwood 1815.

PLATE .II.

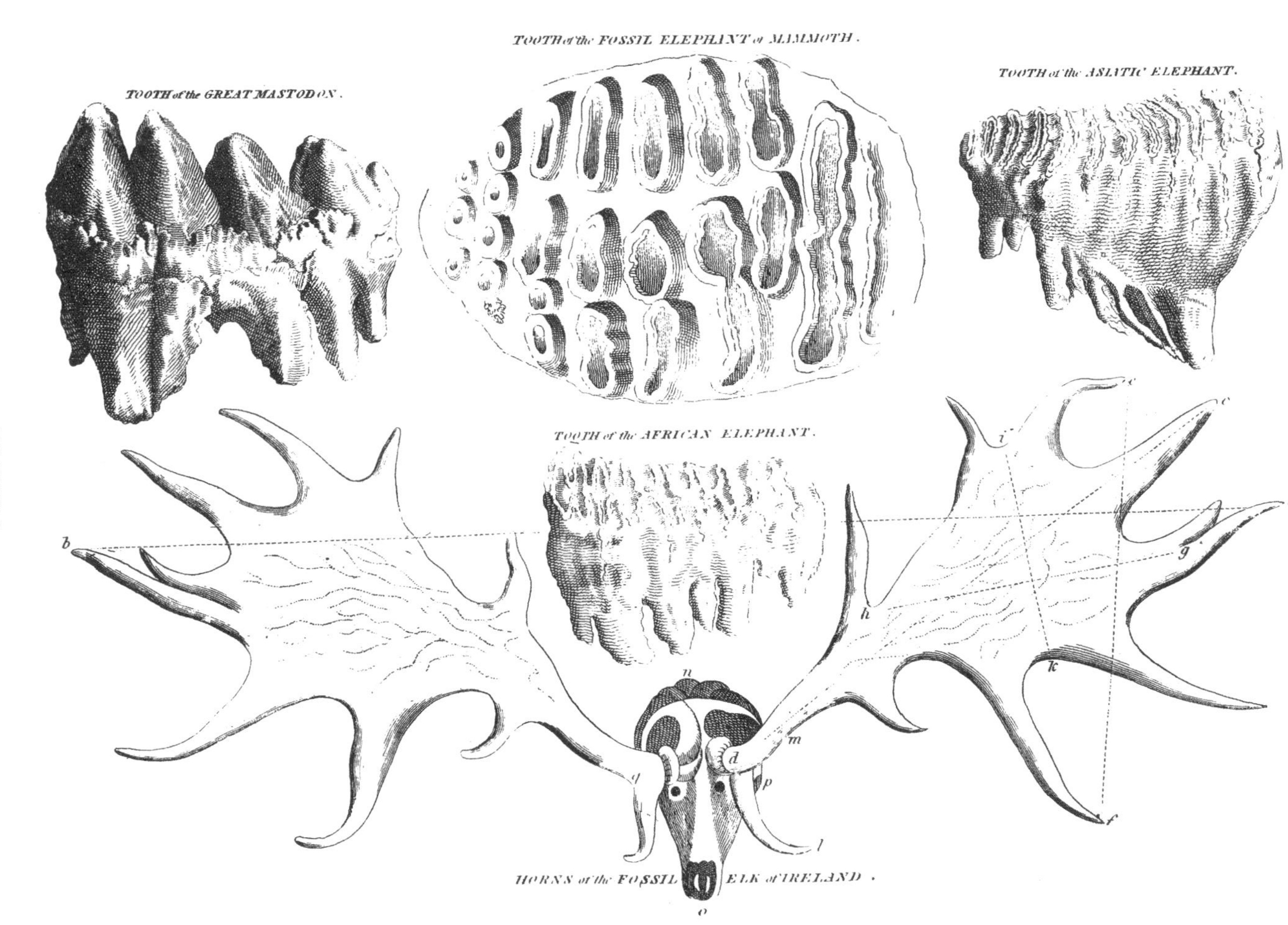

TOOTH of the GREAT MASTODON.
TOOTH of the FOSSIL ELEPHANT or MAMMOTH.
TOOTH of the ASIATIC ELEPHANT.
TOOTH of the AFRICAN ELEPHANT.
HORNS of the FOSSIL ELK of IRELAND.

PLATE III.

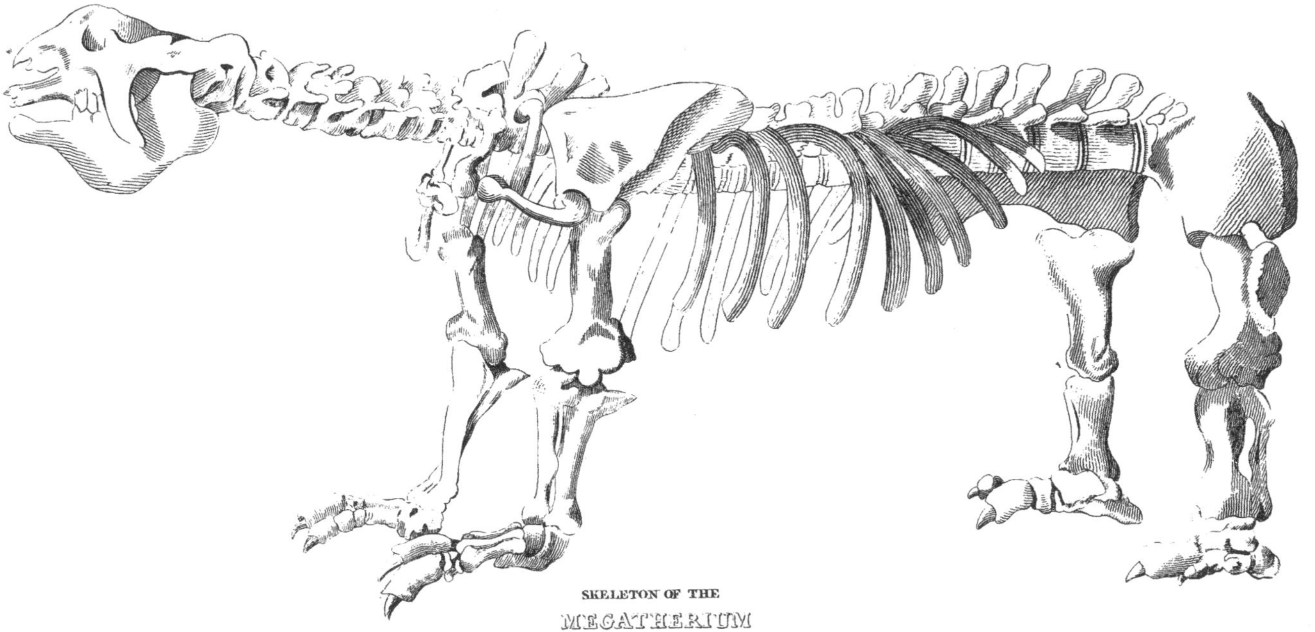

SKELETON OF THE

MEGATHERIUM

DUG OUT of ALLUVIAL STRATA near BUENOS AYRES.

Edinburgh Published by W.m Blackwood 1825.

Eng.d by W. & D. Lizars Edin.

PLATE. IV.

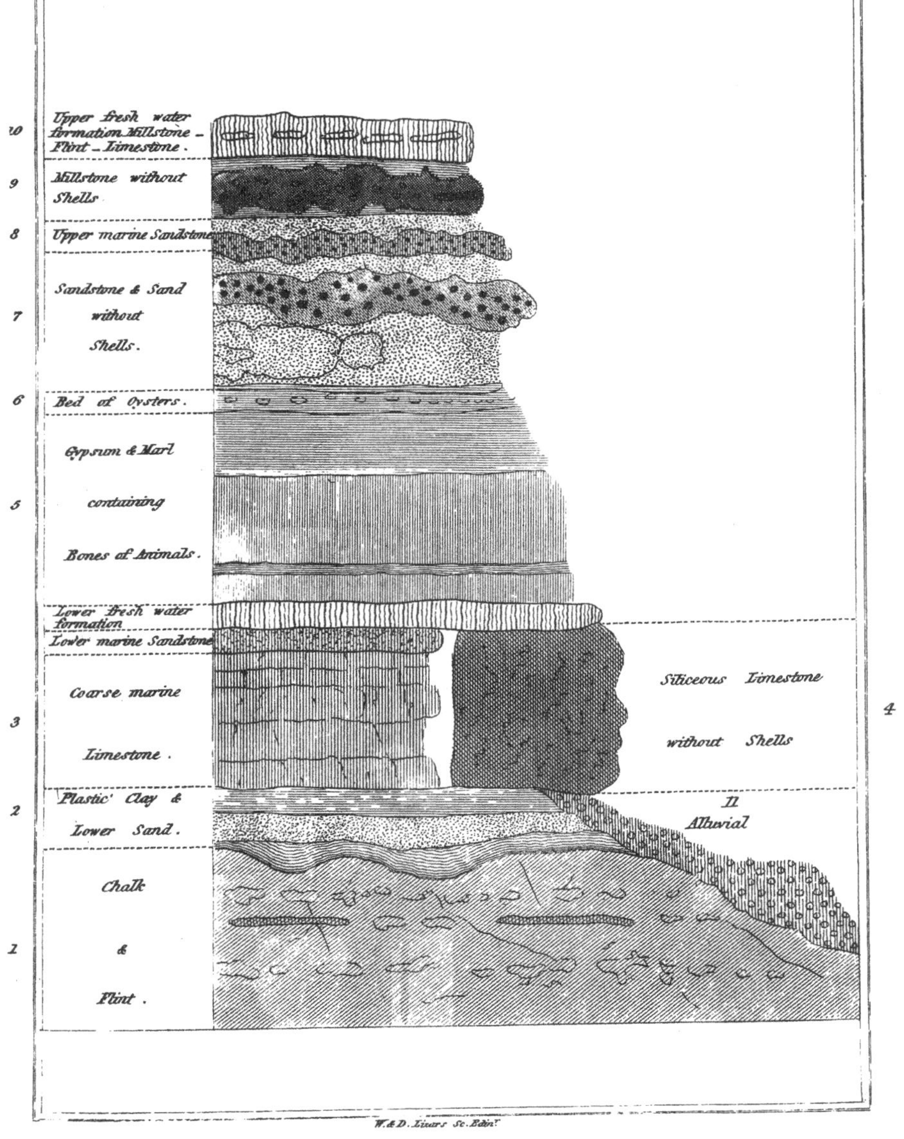

Edinburgh Published by Wm Blackwood 1815.

ESSAY

ON

THE THEORY OF THE EARTH.

§ 1. *Preliminary Observations.*

It is my object, in the following work, to travel over ground which has as yet been little explored, and to make my reader acquainted with a species of Remains, which, though absolutely necessary for understanding the history of the globe, have been hitherto almost uniformly neglected.

As an antiquary of a new order, I have been obliged to learn the art of decyphering and restoring these remains, of discovering and bringing together, in their primitive arrangement, the scattered and mutilated fragments of which they are composed, of reproducing, in all their original proportions and characters, the animals to which these fragments formerly belonged, and then of comparing them with those animals which still live on the surface of the earth; an art which is almost unknown, and which presupposes, what had scarcely been obtained before, an acquaintance with those laws which regulate the coexistence of the forms by which the different parts of organized beings are distinguished. I had next to prepare myself for these enquiries by others of a far more extensive kind, respecting the animals which still exist. Nothing except an almost complete review of creation in its present state, could give a character of demonstration to the results of my investigations into its ancient state; but that review has afforded me, at the same time, a great body of rules and affinities which are no less satisfactorily demonstrated; and the whole animal kingdom has been subjected to new laws in consequence of this Essay on a small part of the theory of the earth.*

---

* This will be seen more at large in the extensive work upon Comparative Anatomy, in which I have been employed for more than twenty-five years, and which I intend soon to prepare for publication.

The importance of the truths which have been developed in the progress of my labours, has contributed equally with the novelty of my principal results to sustain and encourage my efforts. May it have a similar effect on the mind of the reader, and induce him to follow me patiently through the difficult paths in which I am under the necessity of leading him!

The ancient history of the globe, which is the ultimate object of all these researches, is also of itself one of the most curious subjects that can engage the attention of enlightened men; and if they take any interest in examining, in the infancy of our species, the almost obliterated traces of so many nations that have become extinct, they will doubtless take a similar interest in collecting, amidst the darkness which covers the infancy of the globe, the traces of those revolutions which took place anterior to the existence of all nations.

We admire the power by which the human mind has measured the motions of globes which nature seemed to have concealed for ever from our view: Genius and science have burst the limits of space, and a few observations, explained by just reasoning, have unveiled the mechanism of the universe. Would it not also be glorious for man to burst the limits of time, and, by a few observations, to ascertain the history of this world, and the series of events which preceded the birth of the human race? Astronomers, no doubt, have advanced more rapidly than naturalists; and the present period, with respect to the theory of the earth, bears some resemblance to that in which some philosophers thought that the heavens were formed of polished stone, and that the moon was no larger than the Peloponnesus; but, after Anaxagoras, we have had our Copernicuses, and our Keplers, who pointed out the way to Newton; and why should not natural history also have one day its Newton?

§ 2. *Plan of this Essay.*

What I now offer comprehends but a few of the facts which must enter into the composition of this ancient history. But these few are important; many of them are decisive; and I hope that the rigorous methods which I have adopted for the purpose of establishing them, will make them be considered as points so determinately fixed as to admit of no departure from them. Though this hope should only be realised with respect to some of them, I shall think myself sufficiently rewarded for my labour.

In this preliminary discourse I shall describe the whole of the results at which the theory of the earth seems to me to have arrived. I shall mention the relations which connect the history of the fossil bones of lând animals with these results, and the considerations which render their history peculiarly important. I shall unfold the principles on which is founded the art of ascertaining these bones, or, in other words, of discovering a genus and of distinguishing a species by a single fragment of bone,—an art on the certainty of which depends that of the whole work. I shall give a rapid sketch of the results to which my researches lead, of the new species and genera which these have been the means of discovering, and of the different strata in which they are found deposited. And as the difference between these species and the species which still exist is bounded by certain limits, I shall show that these limits are a great deal more extensive than those which now distinguish the varieties of the same species; and shall then point out how far these varieties may be owing to the influence of time, of climate, or of domestication.

In this way I shall be prepared to conclude that great events were necessary to produce the more considerable differences which I have discovered: I shall next take notice of the particular modifications which my performance should introduce into the hitherto received opinions respecting the primitive history of the globe; and, last of all, I shall enquire how far the civil and religious history of different nations corresponds with the results of an examination of the physical history of the earth, and with the probabilities afforded by such examination concerning the period at which societies of men had it in their power to take up fixed abodes, to occupy fields susceptible of cultivation, and consequently to assume a settled and durable form.

§ 3. *Of the first Appearance of the Earth.*

When the traveller passes through those fertile plains where gently flowing streams nourish in their course an abundant vegetation, and where the soil, inhabited by a numerous population, adorned with flourishing villages, opulent cities, and superb monuments, is never disturbed except by the ravages of war and the oppression of tyrants, he is not led to suspect that nature also has had her intestine wars, and that the surface of the globe has been much convulsed by successive revolutions and various catastrophes. But his ideas change as soon as he digs into that soil which presented such a peaceful aspect, or ascends the hills which border the plain; they are expanded, if I may use the expression, in proportion to the expansion of his view; and they begin to embrace the full extent and grandeur of those ancient events to which I have alluded, when he climbs the more elevated chains whose base is skirted by these first hills, or when, by following the beds of the descending torrents, he penetrates into their interior structure, which is thus laid open to his inspection.

§ 4. *First Proofs of Revolutions on the Surface of the Globe.*

The lowest and most level parts of the earth, when penetrated to a very great depth, exhibit nothing but horizontal strata composed of various substances, and containing almost all of them innumerable marine productions. Similar strata, with the same kind of productions, compose the hills even to a great height. Sometimes the shells are so numerous as to constitute the entire body of the stratum. They are almost everywhere in such a perfect state of preservation, that even the smallest of them retain their most delicate parts, their sharpest ridges, and their finest and tenderest processes. They are found in elevations far above the level of every part of the ocean, and in places to which the sea could not be conveyed by any existing cause. They are not only inclosed in loose sand, but are often incrusted and penetrated on all sides by the hardest stones. Every part of the earth, every hemisphere, every continent, every island of any size, exhibits the same phenomenon. We are therefore forcibly led to believe not only that the sea has at one period or another covered all our plains, but that it must have remained there for a long time, and in a state of tranquillity; which circumstance was necessary for the formation of deposits so extensive, so thick, in part so solid, and containing exuviæ so perfectly preserved.

The time is past for ignorance to assert that

these remains of organized bodies are mere *lusus naturæ*,—productions generated in the womb of the earth by its own creative powers. A nice and scrupulous comparison of their forms, of their contexture, and frequently even of their composition, cannot detect the slightest difference between these shells and the shells which still inhabit the sea. They have therefore once lived in the sea, and been deposited by it: the sea consequently must have rested in the places where the deposition has taken place. Hence it is evident that the basin or reservoir containing the sea has undergone some change at least, either in extent, or in situation, or in both. Such is the result of the very first search, and of the most superficial examination.

The traces of revolutions become still more apparent and decisive when we ascend a little higher, and approach nearer to the foot of the great chains of mountains. There are still found many beds of shells; some of these are even larger and more solid; the shells are quite as numerous and as entirely preserved; but they are not of the same species with those which were found in the less elevated regions. The strata which contain them are not so generally horizontal; they have various degrees of inclination, and are sometimes situated vertically. While in the plains and low hills it was necessary to dig deep in order to detect the succession of the strata, here we perceive them by means of the vallies which time or violence has produced, and which disclose their edges to the eye of the observer. At the bottom of these declivities, huge masses of their *debris* are collected, and form round hills, the height of which is augmented by the operation of every thaw and of every storm.

These inclined or vertical strata, which form the ridges of the secondary mountains, do not rest on the horizontal strata of the hills which are situated at their base, and serve as their first steps; but, on the contrary, are situated underneath them. The latter are placed upon the declivities of the former. When we dig through the horizontal strata in the neighbourhood of the inclined strata, the inclined strata are invariably found below. Nay, sometimes, when the inclined strata are not too much elevated, their summit is surmounted by horizontal strata. The inclined strata are therefore more ancient than the horizontal strata. And as they must necessarily have been formed in a horizontal position, they have been subsequently shifted into their inclined or vertical position, and that too before the horizontal strata were placed above them.

Thus the sea, previous to the formation of the horizontal strata, had formed others, which, by some means, have been broken, lifted up, and overturned in a thousand ways. There had therefore been also at least one change in the basin of that sea which preceded ours; it had also experienced at least one revolution; and as several of these inclined strata which it had formed first, are elevated above the level of the horizontal strata which have succeeded and which surround them, this revolution, while it gave them their present inclination, had also caused them to project above the level of the sea, so as to form islands, or at least rocks and inequalities; and this must have happened whether one of their edges was lifted up above the water, or the depression of the opposite edge caused the water to subside. This is the second result, not less obvious, nor less clearly demonstrated, than the first, to every one who will take the trouble of studying carefully the remains by which it is illustrated and proved.

§ 5. *Proofs that such Revolutions have been numerous.*

If we institute a more detailed comparison between the various strata and those remains of animals which they contain, we shall soon discover still more numerous differences among them indicating a proportional number of changes in their condition. The sea has not always deposited stony substances of the same kind. It has observed a regular succession as to the nature of its deposits; the more ancient the strata are, so much the more uniform and extensive are they; and the more recent they are, the more limited are they, and the more variation is observed in them at small distances. Thus the great catastrophes which have produced revolutions in the basin of the sea, were preceded, accompanied, and followed by changes in the nature of the fluid and of the substances which it held in solution; and when the surface of the seas came to be divided by islands and projecting ridges, different changes took place in every separate basin.

Amidst these changes of the general fluid, it

must have been almost impossible for the same kind of animals to continue to live :—nor did they do so in fact. Their species, and even their genera, change with the strata ; and although the same species occasionally recur at small distances, it is generally the case that the shells of the ancient strata have forms peculiar to themselves ; that they gradually disappear, till they are not to be seen at all in the recent strata, still less in the existing seas, in which, indeed, we never discover their corresponding species, and where several even of their genera are not to be found ; that, on the contrary, the shells of the recent strata resemble, as it respects the genus, those which still exist in the sea ; and that in the last-formed and loosest of these strata there are some species which the eye of the most expert naturalist cannot distinguish from those which at present inhabit the ocean.

In animal nature, therefore, there has been a succession of changes corresponding to those which have taken place in the chemical nature of the fluid ; and when the sea last receded from our continent, its inhabitants were not very different from those which it still continues to support.

Finally, if we examine with greater care these remains of organized bodies, we shall discover, in the midst even of the most ancient secondary strata, other strata that are crowded with animal or vegetable productions, which belong to the land and to fresh water; and amongst the more recent strata, that is, the strata which are nearest the surface, there are some of them in which land animals are buried under heaps of marine productions. Thus the various catastrophes of our planet have not only caused the different parts of our continent to rise by degrees from the basin of the sea, but it has also frequently happened, that lands which had been laid dry have been again covered by the water, in consequence either of these lands sinking down below the level of the sea, or of the sea being raised above the level of the lands: The particular portions of the earth also which the sea has abandoned by its last retreat, had been laid dry once before, and had at that time produced quadrupeds, birds, plants, and all kinds of terrestrial productions; it had then been inundated by the sea, which has since retired from it, and left it to be occupied by its own proper inhabitants.

The changes which have taken place in the

productions of the shelly strata, have not, therefore, been entirely owing to a gradual and general retreat of the waters, but to successive irruptions and retreats, the final result of which, however, has been an universal depression of the level of the sea.

§ 6. *Proofs that the Revolutions have been sudden.*

These repeated irruptions and retreats of the sea have neither been slow nor gradual; most of the catastrophes which have occasioned them have been sudden, and this is easily proved, especially with regard to the last of them, the traces of which are most conspicuous. In the northern regions it has left the carcases of some large quadrupeds which the ice had arrested, and which are preserved even to the present day with their skin, their hair, and their flesh. If they had not been frozen as soon as killed they must quickly have been decomposed by putrefaction. But this eternal frost could not have taken possession of the regions which these animals inhabited except by the same cause which destroyed them ;* this cause, therefore, must have been as sudden as its effect. The breaking to pieces and overturnings of the strata, which happened in former catastrophes, shew plainly enough that they were sudden and violent like the last ; and the heaps of *debris* and rounded pebbles which are found in various places among the solid strata, demonstrate the vast force of the motions excited in the mass of waters by these overturnings. Life, therefore, has been often disturbed on this earth by terrible events—calamities which, at their commencement, have perhaps moved and overturned to a great depth the entire outer crust of the globe, but which, since these first commotions, have uniformly acted at a less depth and less generally. Numberless living beings have been the victims of these catastrophes ; some have been destroyed by sudden inundations, others have been laid dry in

---

* The two most remarkable phenomena of this kind, and which must for ever banish all idea of a slow and gradual revolution, are the rhinoceros discovered in 1771 in the banks of the *Vilhoui*, and the elephant recently found by M. Adams near the mouth of the *Lena*. This last retained its flesh and skin, on which was hair of two kinds ; one short, fine, and crisped, resembling wool, and the other like long bristles. The flesh was still in such high preservation, that it was eaten by dogs.

consequence of the bottom of the seas being instantaneously elevated. Their races even have become extinct, and have left no memorial of them except some small fragments which the naturalist can scarcely recognise.

Such are the conclusions which necessarily result from the objects that we meet with at every step of our enquiry, and which we can always verify by examples drawn from almost every country. Every part of the globe bears the impress of these great and terrible events so distinctly, that they must be visible to all who are qualified to read their history in the remains which they have left behind.

But what is still more astonishing and not less certain, there have not been always living creatures on the earth, and it is easy for the observer to discover the period at which animal productions began to be deposited.

§ 7. *Proofs of the Occurrence of Revolutions before the Existence of Living Beings.*

As we ascend to higher points of elevation,

and advance towards the lofty summits of the mountains, the remains of marine animals, that multitude of shells we have spoken of, begin very soon to grow rare, and at length disappear altogether. We arrive at strata of a different nature, which contain no vestige at all of living creatures. Nevertheless their crystallization, and even the nature of their strata, shew that they also have been formed in a fluid; their inclined position and their slopes shew that they also have been moved and overturned; the oblique manner in which they sink under the shelly strata shews that they have been formed before these; and the height to which their bare and rugged tops are elevated above all the shelly strata, shews that their summits have never again been covered by the sea since they were raised up out of its bosom.

Such are those primitive or primordial mountains which traverse our continents in various directions, rising above the clouds, separating the basins of the rivers from one another, serving, by means of their eternal snows, as reservoirs for feeding the springs, and forming in some measure the skeleton, or, as it were, the rough framework of the earth.

The sharp peaks and rugged indentations which mark their summits, and strike the eye at a great distance, are so many proofs of the violent manner in which they have been elevated. Their appearance in this respect is very different from that of the rounded mountains and the hills with flat surfaces, whose recently-formed masses have always remained in the situation in which they were quietly deposited by the sea which last covered them.

These proofs become more obvious as we approach. The vallies have no longer those gently sloping sides, or those alternately salient and re-entrant angles opposite to one another, which seem to indicate the beds of ancient streams. They widen and contract without any general rule; their waters sometimes expand into lakes, and sometimes descend in torrents; and here and there the rocks, suddenly approaching from each side, form transverse dikes, over which the waters fall in cataracts. The shattered strata of these vallies expose their edges on one side, and present on the other side large portions of their surface lying obliquely; they do not correspond in height, but those which on one side form the summit of the declivity, often dip so deep on the other as to be altogether concealed.

Yet, amidst all this confusion, some naturalists have thought that they perceived a certain degree of order prevailing, and that among these immense beds of rock, broken and overturned though they be, a regular succession is observed, which is nearly the same in all the different chains of mountains. According to them, the granite, which surmounts every other rock, also dips under every other rock; and is the most ancient of any that has yet been discovered in the place assigned it by nature. The central ridges of most of the mountain chains are composed of it; slaty rocks, such as clay slate, granular quartz, (*grès*,) and mica slate, rest upon its sides and form lateral chains; granular, foliated limestone, or marble, and other calcareous rocks that do not contain shells, rest upon the slate, forming the exterior ranges, and are the last formations by which this ancient uninhabited sea seems to have prepared itself for the production of its beds of shells.*

* See Pallas, in his Memoir on the Formation of Mountains.

On all occasions, even in districts that lie at a distance from the great mountain chains, where the more recent strata have been digged through, and the external covering of the earth penetrated to a considerable depth, nearly the same order of stratification has been found as that already described. The crystallized marbles never cover the shelly strata; the granite in mass never rests upon the crystallized marble, except in a few places where it seems to have been formed of granites of newer epochs. In one word, the foregoing arrangement appears to be general, and must therefore depend upon general causes, which have on all occasions exerted the same influence from one extremity of the earth to the other.

Hence, it is impossible to deny, that the waters of the sea have formerly, and for a long time, covered these masses of matter which now constitute our highest mountains; and farther, that these waters, during a long time, did not support any living bodies. Thus, it has not been only since the commencement of animal life that these numerous changes and revolutions have taken place in the constitution of the external covering of our globe: For the masses formed previous to that event have suffered changes, as well as those which have been formed since; they have also suffered violent changes in their positions, and a part of these assuredly took place while they existed alone, and before they were covered over by the shelly masses. The proof of this lies in the overturnings, the disruptions, and the fissures which are observable in their strata, as well as in those of more recent formation, which are there even in greater number and better defined.

But these primitive masses have also suffered other revolutions, posterior to the formation of the secondary strata, and have perhaps given rise to, or at least have partaken of, some portion of the revolutions and changes which these latter strata have experienced. There are actually considerable portions of the primitive strata uncovered, although placed in lower situations than many of the secondary strata; and we cannot conceive how it should have so happened, unless the primitive strata, in these places, had forced themselves into view, after the formation of those which are secondary. In some countries, we find numerous and prodigiously large blocks of primitive substances scattered over the surface of the secondary strata, and separated by deep vallies from the peaks or ridges whence these blocks must have been derived. It is necessary, therefore, either that these blocks must have been thrown into those situations by means of eruptions, or that the vallies, which otherwise must have stopped their course, did not exist at the time of their being transported to their present sites. *

Thus we have a collection of facts, a series of epochs anterior to the present time, and of which the successive steps may be ascertained with perfect certainty, although the periods which intervened cannot be determined with any degree of precision. These epochs form so many fixed points, answering as rules for directing our inquiries, respecting this ancient chronology of the earth.

---

* The scientific journies of Saussure and Deluc give a prodigious number of instances of this nature.

§. 8. *Examination of the Causes which act at present on the Surface of our Globe.*

We now propose to examine those changes which still take place on our globe, investigating the causes which continue to operate on its surface, and endeavouring to determine the extent of those effects which they are capable of producing. This portion of the history of the earth is so much the more important, as it has been long considered possible to explain the more ancient revolutions on its surface by means of these still existing causes; in the same manner as it is found easy to explain past events in political history, by an acquaintance with the passions and intrigues of the present day. But we shall presently see that unfortunately this is not the case in physical history: the thread of operation is here broken, the march of nature is changed, and none of the agents that she now employs were sufficient for the production of her ancient works.

There still exist, however, four causes in full activity, which contribute to make alterations on the surface of our earth. These are rains and thaws, which waste down the steep mountains, and occasion their fragments to collect at their bottoms; streams of water, which sweep away these fragments, and afterwards deposit them in places where their current is abated; the sea, which undermines the foundations of elevated coasts, forming steep cliffs in their places, and which throws up hillocks of sand upon flat coasts; and, finally, volcanoes, which pierce through the most solid strata from below, and either elevate or scatter abroad the vast quantity of matter which they eject.

§ 9. *Of Slips, or Falling Down of the Materials of Mountains.*

In every place where broken strata present their edges to the day in abrupt crags, fragments of their materials fall down every spring, and after every storm; these become rounded by rolling upon each other, and their collected heaps assume a determinate inclination or external form, regulated by the laws of cohesion, forming at the bottom of the crag, whence they have fallen, taluses of greater or lesser elevation, in proportion to the quantity of the fragments. These taluses constitute the sides of the vallies in all elevated mountainous regions, and are covered over by abundant vegetation, whenever these fallings down of materials from higher mountains become less frequent; but their want of solidity subjects them also to slips, in consequence of being undermined by the waters of rivulets. On these occasions, towns and rich populous districts are sometimes buried under the ruins of a mountain; the courses of rivers are stopped up, and lakes are formed in places which were before the abodes of fertility and cheerfulness. Fortunately such great slips occur but seldom; and the principal use of these hills, composed of fragments and ruins of the high mountains, is to furnish materials for the ravages of the torrents to operate upon.

§ 10. *Of Alluvial Formations.*

The rains which fall upon the ridges and summits of the mountains, the vapours which are condensed there, and the snow which is melted, descend by an infinite number of rills along their slopes, carrying off some portions of the materials of which these ridges and summits are composed, and marking their courses by numerous gutters. In their progress downwards, these small rills soon unite in the deeper furrows with which the surface of all mountains is ploughed up, run off through the deep vallies which intersect the bottoms of the mountains, and at length form the streams and rivers which restore to the sea the waters that it had formerly supplied to the atmosphere.

When the snow melts, or when a storm takes place, these mountain torrents become suddenly swelled, and rush down the declivities with a violence and rapidity proportioned to their steepness: They dash against the feet of these taluses of fallen fragments which form the sides of all the elevated vallies, carrying along with them the rounded fragments of which they are composed, which become smoothed and still farther polished by rubbing on each other. But, in proportion as the swollen torrents reach the more level vallies, and the force of their current is diminished, or when they arrive at more expanded basins which allow their waters to spread out, they then throw out on their banks the largest of these stones which they had rolled down: The smaller fragments are deposited still lower; and, in general, nothing reaches the great canal of the river except the minutest fragments, or the impalpable particles, which afterwards subside to form mud. It often happens also, before these streams unite to form great rivers, that they have to pass through large and deep lakes, where they deposit the mud brought down from the mountains, and whence their waters flow out quite limpid.

The rivers in lower levels, and all the streams which take their rise in lower mountains or hills, produce effects on the grounds through which they flow, more or less analogous to those of the torrents from the higher mountains. When swelled by great rains, they undermine the bottoms of the earthy or sandy hills which lie in their way, and carry their fragments to be deposited on the lower grounds which they inundate, and which are somewhat raised in height by each successive inundation. Finally, when these rivers reach the great lakes, or the sea, and when of course that rapid motion by which they are enabled to keep the particles of mud in suspension has wholly ceased, these particles are deposited at each side of their mouths, where they form low grouuds, by which the coasts or banks of the river. are gradually lengthened out into the sea or lake. And if these new coasts are so situated that the sea also throws up sand to contribute towards their increase, provinces, and even entire kingdoms, are thus as it were created, which usually become the richest and most fertile regions, if their rulers permit human industry to exert itself in peace.

§ 11. *Of the Formation of Downs.*

The effects produced by the sea alone, without the aid of rivers, are far less beneficial. When the sea coast is low, and the bottom consists of sand, the waves push this sand towards the shore, where at every reflux of the tide it becomes partially dried; and the winds, which almost always blow from the sea, drift up some portion of it upon the beach. By this means, *downs*, or ranges of low sand hills, are formed along the coast. These, if not fixed by the growth of suitable plants, either disseminated by nature, or propagated by human industry, would be gradually, but certainly carried towards the interior, covering up the fertile plains with their sterile particles, and rendering them unfit for the habitation of mankind; because the same winds which carried the loose dry sand from the shore to form the downs, would necessarily continue to drift that which is at the summit farther towards the land.

§ 12. *Of the Formation of Cliffs, or steep Shores.*

On the other hand, when the original coast happens to be high, so that the sea is unable to cast up any thing upon it, a gradual, but destructive operation is carried on in a different way. The incessant agitation of the waves wears it away at the bottom, and at length succeeds in undermining it, causing the upper materials to slide and tumble down, and converting the whole elevation into steep sloping bluffs or cliffs. In the progress of this change, the more elevated materials which tumble down into the sea, have their softer parts washed out and carried away by the waves; while the harder parts, continually rolled about in the agitated water, form vast collections of rounded stones and pebbles, and of sand of various degrees of fineness, which at length accumulate into sloping banks or flat beaches, and protect the bottoms of the cliffs against farther depredations.

Such are the ordinary actions of water upon the solid land, which almost entirely consist in reducing it to lower levels, but not indefinitely. The fragments of the great mountain ridges are carried down into the vallies, while their finer particles and those of the lower hills and plains are floated to the sea. Alluvial depositions extend the coast at the expence of the interior hills, which last effect is mostly limited in its extent by means of vegetation. All these changes necessarily suppose the previous existence of mountains, vallies, and plains, and consequently the same causes could not have given rise to these inequalities on the surface of our globe.

The formation of downs is the most limited of all these phenomena, both in regard to height and horizontal extent, and has no manner of relation whatever to those enormous masses, the origin of which forms the peculiar object of geological research.

§ 13. *Of Depositions formed in Water.*

Although we cannot obtain a precise knowledge of the actions exerted by water within its own bosom, still it may be ascertained in a certain degree.

Lakes, low meadows, marshes, and sea-ports, into which rivulets discharge their waters, more especially when these descend from near and steep hills, are continually receiving depositions of mud, which would at length fill them up entirely, if they were not carefully cleaned out. The sea is constantly accumulating quantities of sand and slime into its bays and harbours, or wherever its waters happen to become more quiet than ordinary. The currents also occasioned by the tides, are continually washing large quantities of sand from the bottom of the sea, which they collect together and heap up on various parts of the coast, forming banks and flat shallows.

§ 14. *Of Stalactites.*

Certain waters, after dissolving calcareous substances by means of the superabundant carbonic acid with which they are impregnated, allow these substances to crystallize, in consequence of the escape of the acid, and in this way form stalactites and other concretions. There are some strata, confusedly crystallized in fresh water, which are sufficiently extensive to be compared with other strata that have been left by the ancient sea.

§ 15. *Of Lithophites.*

In the torrid zone, where lithophites of many kinds abound, and are propagated with great rapidity, their stony tree-like fabrics are intertwined and accumulated into the form of rocks and reefs, and, rising even to the surface of the water, shut up the entrance of harbours, and lay frightful snares for navigators. The sea, throwing up sand and mud on the tops of those rocky shelves, sometimes raises them above its own proper level, and forms islands of them, which are soon covered with a rich vegetation.

§ 16. *Of Incrustations.*

It is also possible that the animals inhabiting shells may leave their stony coverings when they die in some particular places; and that these, cemented together by slime of greater or less consistence, or by some other means, may form extensive banks of shells. But we have no evidence that the sea has now the power of agglutinating these shells by such a compact paste, or indurated cement, as that found in marbles and calcareous sand-stones, or even in the coarse limestone strata in which shells are found enveloped. Still less do we now find the sea making any depositions at all of the more solid and silicious strata which have preceded the formation of the strata containing shells. In short, all these causes would not, though combined, form a single stratum of any kind, nor produce the smallest hillock, nor alter in any perceptible degree the ordinary level of the ocean.

It has been asserted that the sea is subject to

a continual diminution in its level; and proofs of this are said to have been discovered in some parts of the shores of the Baltic. Whatever may have been the cause of these appearances, we certainly know that nothing of the kind has been observed upon our coasts; and, consequently, that there has been no general lowering of the waters of the ocean. The most ancient sea-ports still have their quays and other erections at the same height above the level of the sea as at their first construction.

Certain general movements have been supposed in the sea from east to west, or in other directions; but no where has any person been able to ascertain their effects with the least degree of precision.

§ 17. *Of Volcanoes.*

The operation of volcanoes is still more limited and local than that of any of the agents which have yet been mentioned. Although we have no idea of the means employed by nature for feeding these enormous fires from such vast depths, we can judge decidedly by their effects.

of the changes which they were capable of producing upon the surface of the earth. When a volcano announces itself after some shocks of an earthquake, it forms for itself an opening. Stones and ashes are thrown to a great distance, and lava is vomited forth. The more fluid part of the lava runs in long streams, while the less fluid portion stops at the edge of the opening, raises it all round, and forms a cone terminated by a crater. Thus volcanoes accumulate substances on the surface that were formerly buried deep in the bowels of the earth, after having changed or modified their nature or appearances, and raise them into mountains. By these means, they have formerly covered some parts of the continents, and have suddenly produced mountains in the middle of the sea. But these mountains and islands have always been composed of lava, and the whole of their materials have undergone the action of fire. Volcanoes have never raised up nor overturned the strata through which their apertures pass, and have in no degree contributed to the elevation of the great mountains which are not volcanic.

Thus we shall seek in vain among the various forces which still operate on the surface of our earth, for causes competent to the production of those revolutions and catastrophes of which its external crust exhibits so many traces: And, if we have recourse to the constant external causes with which we have been hitherto acquainted, we shall have no greater success.

§ 18. *Of Astronomical Causes of the Revolutions on the Surface of the Earth.*

The pole of the earth moves in a circle round the pole of the ecliptic, and its axis is more or less inclined to the plane of the ecliptic; but these two motions, the causes of which are now ascertained, are confined within certain bounds, and are much too limited for the production of those effects which we have stated. Besides, as these motions are exceedingly slow, they are altogether inadequate to account for catastrophes which must necessarily have been sudden.

The same reasoning applies to all other slow motions which have been conceived as causes of the revolutions on the surface of our earth, chosen doubtless in the hope that their existence could not be denied, as it might always be asserted that their extreme slowness rendered them imperceptible. But it is of no importance whether these assumed slow motions be true or false, for they explain nothing, since no cause acting slowly could possibly have produced sudden effects.

Admitting that there was a gradual diminution of the waters; that the sea might take away solid matters from one place and carry them to another; that the temperature of the globe may have diminished or increased; none of these causes could have overthrown our strata; inclosed great quadrupeds with their flesh and skin in ice; laid dry sea-shells in as perfect preservation as if just drawn up alive from the bottom of the ocean; or utterly destroyed many species, and even entire genera, of testaceous animals.

These considerations have presented themselves to most naturalists: And, among those who have endeavoured to explain the present state of the globe, hardly any one has attributed the entire changes it has undergone to slowly operating causes, and still less to causes which continue to act, as it were, under our observation. The necessity to which they were thus reduced, of seeking for causes different from those which we still observe in activity, is the very thing which has forced them to make so many extraordinary suppositions, and to lose themselves in so many erroneous and contradictory speculations, that the very name of their science, as I have elsewhere said, has become ridiculous in the opinion of prejudiced persons, who only see in it the systems which it has exploded, and forget the extensive and important series of facts which it has brought to light and established. *

§ 19. *Of former Systems of Geology.*

During a long time, two events or epochs only, the Creation and the Deluge, were admitted as comprehending the changes which have occurred upon the globe; and all the efforts of geologists were directed to account for the present actual

* When I formerly mentioned this circumstance, of the science of geology having become ridiculous, I only expressed a well known truth, without presuming to give my own opinion, as some respectable geologists seem to have believed. If their mistake arose from my expressions having been rather equivocal, I take this opportunity of explaining my meaning.

state of the earth, by arbitrarily ascribing to it a certain primitive state, afterwards changed and modified by the deluge, of which also, as to its causes, its operation, and its effects, every one of them entertained his own theory.

Thus, in the opinion of *Burnet*,* the whole earth at the first consisted of a uniform light crust, which covered over the abyss of the sea, and which, being broken for the production of the deluge, formed the mountains by its fragments. According to *Woodward*,† the deluge was occasioned by a momentary suspension of cohesion among the particles of mineral bodies; the whole mass of the globe was dissolved, and the soft paste became penetrated by shells. *Scheuchzer*‡ conceived that God raised up the mountains for the purpose of allowing the waters of the deluge to run off, and accordingly selected those portions which contained the greatest abundance of rocks, without which they could

---

* Telluris Theoria Sacra. Lond. 1681.

† Essay towards the Natural History of the Earth. Lond. 1702.

‡ Memoires de l'Academie, 1708.

not have supported themselves. *Whiston* * fancied that the earth was created from the atmosphere of one comet, and that it was deluged by the tail of another. The heat which remained from its first origin, in his opinion, excited the whole antediluvian population, men and animals, to sin, for which they were all drowned in the deluge, excepting the fish, whose passions were apparently less violent.

It is easy to see, that though naturalists might have a range sufficiently wide within the limits prescribed by the book of Genesis, they very soon found themselves in too narrow bounds; and when they had succeeded in converting the six days employed in the work of creation into so many periods of indefinite length, their systems took a flight proportioned to the periods, which they could then dispose of at pleasure.

Even the great *Leibnitz*, as well as *Descartes*, amused his imagination by conceiving the world to be an extinguished sun, or vitrified globe; upon which the vapours condensing in propor-

* A New Theory of the Earth. Lond. 1708.

tion as it cooled, formed the seas, and afterwards deposited calcareous strata. *

By *Demaillet*, the globe was conceived to have been covered with water for many thousand years. He supposed that this water had gradually retired; that all the terrestrial animals were originally inhabitants of the sea; that man himself began his career as a fish: And he asserts, that it is not uncommon, even now, to meet with fishes in the ocean, which are still only half men, but whose descendants will in time become perfect human beings. †

The system of *Buffon* is merely an extension of that before devised by Leibnitz, with the addition only of a comet, which, by a violent blow upon the sun, struck off the mass of our earth in a liquified state, along with the masses of all the other planets of our system at the same instant. From this supposition, he was enabled to assume positive dates or epochs: As, from the actual temperature of the earth, it could be calculated

---

* Leibnitz, Protogœa. *Act. Lips.* 1683; *Gott.* 1749.
† Telliamed.

how long time it had taken to cool so far. And, as all the other planets had come from the sun at the same time, it could also be calculated how many ages were still required for cooling the greater ones, and how far the smaller ones were already frozen.

In the present day, men of bolder imaginations than ever, have employed themselves on this great subject. Some writers have revived and greatly extended the ideas of Demaillet. They suppose that every thing was originally fluid; that this universal fluid gave existence to animals, which were at first of the simplest kind, such as the monads and other infusory microscopic animalcules; that, in process of time, and by acquiring different habits, the races of these animals became complicated, and assumed that diversity of nature and character in which they now exist. It is by all those races of animals that the waters of the ocean have been gradually converted into calcareous earth; while the vegetables, concerning the origin and metamorphoses of which these authors give us no account, have converted a part of the same water into clay; and these two earths, after being stript of the peculiar characters they had received respectively from animal and vegetable life, are resolved by a final analysis into silex: Hence the more ancient mountains are more silicious than the rest. Thus, according to these authors, all the solid particles of our globe owe their existence to animal or vegetable life, and without this our globe would still have continued entirely liquid. *

Other writers have preferred the ideas of Kepler, and, like that great astronomer, have considered the globe itself as possessed of living faculties. According to them, it contains a circulating vital fluid. A process of assimilation goes on in it as well as in animated bodies. Every particle of it is alive. It possesses instinct and volition even to the most elementary of its molecules, which attract and repel each other according to sympathies and antipathies. Each kind of mineral substance is capable of converting immense masses of matter into its own peculiar nature, as we convert our aliment into flesh and blood. The mountains are the respiratory

---

* See La Physique de Rodig. p. 106. Leipsic, 1801, and Telliamed, p. 169. Lamarck has expanded this system at great length, and supported it with much sagacity, in his *Hydrogéologie*, and *Philosophie Zoologique*.

organs of the globe, and the schists its organs of secretion. By the latter it decomposes the waters of the sea, in order to produce volcanic eruptions. The veins in strata are caries, or abscesses of the mineral kingdom, and the metals are products of rottenness and disease, to which it is owing that almost all of them have so bad a smell.*

It must, however, be noticed, that these are what may be termed extreme examples, and that all geologists have not permitted themselves to be carried away by such bold or extravagant conceptions as those we have just cited. Yet, among those who have proceeded with more caution, and have not searched for geological causes beyond the established limits of physical and chemical science, there still remain much diversity and contradiction.

According to one of these writers, every thing has been successively precipitated and deposited,

---

* M. Patrin has used much ingenuity to establish this view of the subject, in several articles of the *Nouveau Dictionnaire d'Histoire Naturelle.*

nearly as it exists at present; but the sea, which covered all, has gradually retired. *

Another conceives, that the materials of the mountains are incessantly wasted and floated down by the rivers, and carried to the bottom of the ocean, to be there heated under an enormous pressure, and to form strata which shall be violently lifted up at some future period, by the heat that now consolidates and hardens them. †

A third supposes the fluid materials of the globe to have been divided among a multitude of successive lakes, placed like the benches of an amphitheatre; which, after having deposited our shelly strata, have successively broken their dikes, to descend and fill the basin of the ocean. ‡

According to a fourth, tides of seven or eight hundred fathoms have carried off from time to

---

* In his Geology, Delametherie assumes crystallization as the chief cause or agent.

† Hutton, and Playfair in his Illustrations of the Huttonian Theory of the Earth. *Edinb.* 1802.

‡ See Lamanon, in various parts of the Journal de Physique.

time the bottom of the ocean, throwing it up in mountains and hills on the primitive vallies and plains of the continent. *

A fifth conceives the various fragments of which the surface of the earth is composed to have fallen successively from heaven, in the manner of meteoric stones, and alleges that they still retain the marks of their origin in the unknown species of animals whose exuviæ they contain. †

By a sixth, the globe is supposed to be hollow, and to contain in its cavity a nucleus of loadstone, which is dragged from one pole of the earth to the other by the attraction of comets, changing the centre of gravity, and consequently hurrying the great body of the ocean along with it, so as alternately to drown the two hemispheres.‡

---

* Dolomieu, in the Journal de Physique.

† M. M. de Marschall, in Researches respecting the Origin and Developement of the present State of the Earth. *Geissen*, 1802.

‡ Bertrand, Periodical Renewal of the Terrestrial Continents. *Hamburgh*, 1799.

§ 20. *Diversities of the Geological Systems, and their Causes.*

We might have cited twenty other systems, as different from one another as these just now enumerated. And to prevent mistake, we wish it to be distinctly understood, that it is by no means our intention to criticise their authors; on the contrary, we are ready to admit that these systems have generally been conceived by men of science and genius, none of whom were ignorant of the facts on which they reasoned, and several of whom had made extensive journies for the purpose of examining them.

Whence comes it then, that there should be so much contrariety in the solutions of the same problem, that are given by men who proceed upon the same principles? This may have been occasioned by the conditions of the problem never having been all taken into consideration; by which it has remained hitherto indeterminate, and susceptible of many solutions—all equally good, when such or such conditions are abstracted; and all equally bad, when a new condition comes to be known, or when the attention is directed to some known condition, which had been formerly neglected.

§ 21. *Statement of the Nature and Conditions of the Problem to be solved.*

To quit the language of mathematics, it may be asserted, that almost all the authors of these systems, confining their attention to certain difficulties by which they were struck more forcibly than by others, have endeavoured to solve these in a way more or less probable, and have allowed others to remain unnoticed, equally numerous and equally important. For example, the only difficulty with one consisted in explaining the change which had taken place on the level of the seas; with another it consisted in accounting for the solution of all terrestrial substances in the same fluid; and with a third, it consisted in shewing how animals that were natives of the torrid, could live under the frigid zone. Exhausting the whole of their ingenuity on these questions, they conceived that they had done every thing that was necessary, when they had contrived some method of answering them; and yet, while they neglected all the other phenomena, they did not always think of determining with precision the measure and extent of those which they attempted to explain. This is peculiarly the case in regard to the secondary stratifications, which constitute, however, the most difficult and most important portion of the problem. It has hardly ever been attempted carefully to ascertain the superpositions of their strata, or the connections of these strata with the species of animals and of plants whose remains they inclose.

Are there certain animals and plants peculiar to certain strata, and not found in others? What are the species that appear first in order, and those which succeed? Do these two kinds of species ever accompany one another? Are there alternations in their appearances; or in other words, does the first species appear a second time, and does the second species then disappear? Have these animals and plants lived in the places where their exuviæ are found, or have they been brought there from other places? Do all these animals and plants still continue to live in some part of the earth, or have they been totally or partially destroyed? Is there any constant connection between the antiquity of the strata, and the resemblance or non-resemblance of the extraneous fossils; to the animals and plants that still exist? Is there any connection in regard to climate, between the extraneous fossils and the still living organized bodies which most nearly resemble them? May it be concluded, that the transportation of these living organized bodies, if such a thing ever happened, has taken place from north to south, or from east to west; or was it effected by means that irregularly scattered and mingled them together? And, finally, is it still possible to distinguish the epochs of these transportations, by attentively examining the strata which inclose the remains, or are imprinted by their forms?

If, from the want of sufficient evidence, these questions cannot be satisfactorily answered, how shall we be able to explain the causes of the presently existing state of our globe? It is certain, that so far from any of these points being as yet completely established, naturalists seem to have scarcely any idea of the propriety of investigating facts before they construct their systems. The cause of this strange procedure may be discovered, by considering that all geologists hitherto have either been mere cabinet naturalists, who had themselves hardly paid any attention to the structure of mountains, or mere mineralogists, who had not studied in sufficient detail the innumerable diversity of animals, and the almost infinite complication of their various parts and organs. The former of these have only constructed systems; while the latter have made excellent collections of observations, and have laid the foundations of true geological science, but have been unable to raise and complete the edifice.

§ 22. *Of the Progress of Mineral Geology.*

The purely mineralogical portion of the great problem of the Theory of the Earth has been investigated with admirable care by Saussure, and has been since explained in an astonishing degree by Werner, and by the numerous enlightened pupils of his school.

The former of these celebrated philosophers, by a laborious investigation of the most inaccessible mountain districts during twenty years of continual research, in which he examined the Alps on all sides, and penetrated through all their defiles, has laid open to our view the entire disorder of the primitive formations, and has clearly traced the boundaries by which they are distinguishable from the secondary formations. The other equally celebrated geologist, taking advantage of the numerous excavations in the most ancient mining district in the world, has fixed the laws which regulate the succession of strata, pointing out their respective antiquity in regard to each other, and tracing each of them through all its changes and metamorphoses. From him alone we date the commencement of real geology, so far as respects the mineral natures of the strata: But neither he nor Saussure has defined the species of organized extraneous fossils in each description of the strata with that accuracy which has become necessary, now that the number of animals already known has become so great.

Other naturalists, it is true, have studied the fossil remains of organized bodies; they have collected and represented them by thousands, and their works certainly will serve as a valuable storehouse of materials. But, considering these fossil plants and animals merely in themselves, instead of viewing them in their connection with the theory of the earth; or regarding their petrifactions and extraneous fossils as mere curiosities, rather than as historical documents; or confining themselves to partial explanations of the particular bearings of each individual specimen; they have almost always neglected to investigate the general laws affecting their position, or the relation of the extraneous fossils with the strata in which they are found.

§ 23. *Of the Importance of Extraneous Fossils, or Petrifactions, in Geology.*

The importance of investigating the relations of extraneous fossils with the strata in which they are contained, is quite obvious. It is to them alone that we owe the commencement even of a Theory of the Earth; as, but for them, we could never have even suspected that there had existed any successive epochs in the formation of our earth, and a series of different and consecutive operations in reducing it to its present state. By them alone we are enabled to ascertain, with the utmost certainty, that our earth has not always been covered over by the same external crust; because we are thoroughly assured that the organized bodies to which these fossil remains belong, must have lived upon the surface, before they came to be buried, as they now are, at a great depth. It is only by means of analogy, that we have been enabled to extend to the primitive formations, the same conclusions which are furnished directly for the secondary formations by the extraneous fossils; and if there had only existed formations or strata in which there were no extraneous fossils, it could never have been asserted that these several formations had not been simultaneous.

It is also owing to these extraneous fossils, slight as is the knowledge we have hitherto acquired respecting them, that we have been enabled to discover the little that we yet know concerning the revolutions of our globe. From them we have learned, that the strata, or at least those which contain their remains, have been quietly deposited in a fluid; that the variations of the several strata must have corresponded with the variations in the nature of the fluid; that they have been left bare by the transportation of this fluid to some other place; and that this fact must have happened more than once. Nothing of all this could have been known with certainty, without the aid of extraneous fossils.

The study of the mineralogical part of geology, though not less necessary, and even a great deal more useful to the practical arts, is yet much less instructive so far as respects the objects of our present inquiry. We remain in utter ignorance respecting the causes which have given rise to the variety in the mineral substances of which the strata are composed. We are ignorant even of the agents which may have held some of these substances in a state of solution; and it is still disputed respecting several of them, whether they have owed their origin to the agency of water or of fire. After all, philosophers are only agreed on one point, which is, that the sea has changed its place; and this could never have been certainly known, but for the existence of extraneous fossils. These fossils, then, which have given rise to the theory of the earth, have at the same time furnished its principal illustrations—the only ones, indeed, that have as yet been generally received and acknowledged.

This is the consideration by which I have been encouraged to investigate the subject of extraneous fossils. But the field is extensive; and it is only a very inconsiderable portion of it that can be cultivated by the labour of a single individual. It was necessary, therefore, to select a particular department, and I very soon made my choice. That class of extraneous fossils, which forms the peculiar subject of this Essay, engaged my attention at the very outset, because it is evidently the most fertile in affording precise results, yet at the same time less known than others, and richer in new objects of research. *

§ 24. *High Importance of investigating the Fossil Remains of Quadrupeds.*

It is obvious that the fossil remains of the bones of quadrupeds must lead to more rigorous conclusions than any other remains of organized bodies, and that for several reasons.

In the first place, they indicate much more clearly the nature of the revolutions to which

* My work on this subject will clearly show how far this enquiry is yet new, notwithstanding the excellent labours of Camper, Pallas, Blumenbach, Merk, Sœmmerring, Rosenmuller, Fischer, Faujus, and other learned men, whose works I have most scrupulously cited in such of my chapters as their researches are connected with.

they have been subjected. The remains of shells certainly indicate that the sea has once existed in the places where these collections have been formed : But the changes which have taken place in their species, when rigorously enquired into, may possibly have been occasioned by slight changes in the nature of the fluid in which they were formed, or only in its temperature, and may even have arisen from other accidental causes. We can never be perfectly assured that certain species, and even genera, inhabiting the bottom of the sea, and occupying certain fixed spaces for a longer or shorter time, may not have been driven away from these by other species or genera.

In regard to quadrupeds, on the contrary, every thing is precise. The appearance of their bones in strata, and still more of their entire carcases, clearly establishes that the bed in which they are found must have been previously laid dry, or at least that dry land must have existed in its immediate neighbourhood. Their disappearance as certainly announces that this stratum must have been inundated, or that the dry land had ceased to exist in that state. It is from them, therefore, that we learn with perfect certainty the important fact of the repeated irruptions of the sea upon the land, which the extraneous fossils and other productions of marine origin could not of themselves have proved; and, by a careful investigation of them, we may hope to ascertain the number and the epochs of those irruptions of the sea.

Secondly, the nature of the revolutions which have changed the surface of our earth, must have exerted a more powerful action upon terrestrial quadrupeds than upon marine animals. As these revolutions have consisted chiefly in changes of the bed of the sea, and as the waters must have destroyed all the quadrupeds which they reached, if their irruption over the land was general, they must have destroyed the entire class, or, if confined only to certain continents at one time, they must have destroyed at least all the species inhabiting these continents, without having the same effect upon the marine animals. On the other hand, millions of aquatic animals may have been left quite dry, or buried in newly-formed strata, or thrown violently on the coasts, while their races may have been still preserved in more peaceful parts of the sea, whence they might again propagate and spread after the agitation of the water had ceased.

Thirdly, this more complete action is also more easily ascertained and demonstrated; because, as the number of terrestrial quadrupeds is limited, and as most of their species, at least the large ones, are well known, we can more easily determine whether fossil bones belong to a species which still exists, or to one that is now lost. As, on the other hand, we are still very far from being acquainted with all the testaceous animals and fishes belonging to the sea, and as we probably still remain ignorant of the greater part of those which live in the extensive deeps of the ocean, it is impossible to know, with any certainty, whether a species found in a fossil state may not still exist somewhere alive. Hence some naturalists persist in giving the name of oceanic or pelagic shells to *belemnites* and *cornua-ammonis*, and some other genera, which have not hitherto been found, except in the fossil state, in ancient strata; meaning by this, that although these have not as yet been found in a living or recent state, it is because they inhabit the bottom of the ocean, far beyond the reach of our nets.

§ 25. *Of the small probability of discovering new Species of the larger Quadrupeds.*

Naturalists certainly have neither explored all the continents, nor do they as yet know even all the quadrupeds of those parts which have been explored. New species of this class are discovered from time to time; and those who have not examined with attention all the circumstances belonging to these discoveries, may allege also that the unknown quadrupeds, whose fossil bones have been found in the strata of the earth, have hitherto remained concealed in some islands not yet discovered by navigators, or in some of the vast deserts which occupy the middle of Africa, Asia, the two Americas, and New Holland. But, if we carefully attend to the kinds of quadrupeds that have been recently discovered, and to the circumstances of their discovery, we shall easily perceive that there is very little chance indeed of our ever finding alive those which have only been seen in a fossil state.

Islands of moderate size, and at a considerable distance from the large continents, have very few quadrupeds, and these mostly very small.

When they contain any of the larger quadrupeds, these must have been carried to them from other countries. Cook and Bougainville ſound no other quadrupeds besides hogs and dogs in the South Sea islands; and the largest quadruped of the West India islands, when first discovered, was the *agouti*, a species of the *cavy*, an animal apparently between the rat and the rabbit.

It is true that the great continents, as Asia, Africa, the two Americas, and New Holland, have large quadrupeds, and, generally speaking, contain species proper to each: Insomuch, that, upon discovering countries which are isolated from the rest of the world, the animals they contain of the class of quadrupeds were found entirely different from those which existed in other countries. Thus, when the Spaniards first penetrated into South America, they did not find it to contain a single quadruped exactly the same with those of Europe, Asia, and Africa. The puma, the jaguar, the tapir, the capybara, the lama, or glama, the vicugna, and the whole tribe of sapajous, were to them entirely new animals, of which they had not the smallest idea.

Similar circumstances have recurred in our

own time, when the coasts of New Holland and the adjacent islands were first examined. The species of the kangaroo, *phascoloma*, *dasyurus*, *peramela*, *phalanger*, or flying opposum, with the hairy and spinous duck-billed animals denominated *ornithorinchus* and *echidna*, * have astonished zoologists by presenting new and strange conformations, contrary to all former rules, and incapable of being reduced under any of the former systems.

If there still remained any great continent to be discovered, we might perhaps expect to be made acquainted with new species of large quadrupeds; among which some might be found more or less similar to those of which we find the exuviæ in the bowels of the earth. But it is merely sufficient to glance the eye over the map of the world, and observe the innumerable directions in which navigators have traversed the ocean, in order to be satisfied that there does not

* These are new animals of Australia or new Holland, only recently discovered, whose strange conformations, not analogous with the animals of the old world, or of America, have required the adoption of new generic terms by Cuvier and other naturalists.—*Transl.*

remain any large land to be discovered, unless it may be situated towards the antarctic pole, where eternal ice necessarily forbids the existence of animal life.

Hence, it is only from the interiors of the large divisions of the world already known, that we can now hope to procure any quadrupeds hitherto unknown. But a very little reflection will be sufficient to convince us, that our hopes from thence are not much better founded than from the larger islands.

Doubtless, European travellers cannot easily penetrate through vast extents of countries which are either uninhabited, or peopled only with ferocious tribes; and this is peculiarly the case in regard to Africa. But there is nothing to prevent the animals themselves from roaming in all directions, and penetrating to the coasts. Even although great chains of mountains may intervene between the coasts and the interior deserts, these must certainly be broken in some parts, to allow the rivers to pass through; and in these burning deserts the animals naturally follow the courses of the rivers. The inhabitants of the coast must also frequently penetrate inland along the rivers, and will quickly acquire a knowledge of all the remarkable living creatures, even to the very sources of these rivers, either from personal observation, or by intercourse with the inhabitants of the interior. At no period of our history, therefore, could civilised nations frequent the coasts of large countries for any length of time, without gaining some tolerable knowledge of all the animals they contained, or at least of such as were any way remarkable for their size or configuration. This reasoning is supported by well-known facts. Thus, although the ancients seem never to have passed the mountains of Imaus, or to have crossed the Ganges towards the east of Asia, and never penetrated far to the south of Mount Atlas in Africa, yet they were acquainted with all the larger animals of these two grand divisions of the world ; and if they have not distinguished all their species, it was because the similarities of some of these occasioned them to be confounded together, and not because they had not seen them, or heard them talked of by others.

The ancients were perfectly acquainted with the elephant, and the history of that quadruped is given more exactly by Aristotle than by Buffon. They were not ignorant even of the differences which distinguish the elephants of Africa from those of Asia. *

They knew the two-horned rhinoceros, which Domitian exhibited in his shews at Rome, and had stamped on his medals, and of which Pausanias has left a very good description. Even the one-horned rhinoceros, although its country be far from Rome, was equally known to the Romans; Pompey shewed them one in the circus, and Strabo has described another which he saw at Alexandria. †

The hippopotamus has not been so well described by the ancients as the two foregoing animals; yet very exact representations of it have been left by the Romans in their monuments relative to Egypt, such as the statue of the Nile, the Prenestine pavement, and a great number of medals. It is known that this animal was frequently shewn to the Romans, having been ex-

---

* See this more particularly noticed in the history of the elephant, in the second volume of my Researches into the Extraneous or Fossil Remains of Quadrupeds.

† See the history of the Rhinoceros in my second volume.

hibited in the circus by Scaurus, Augustus, Antoninus, Commodus, Heliogabalus, Philip,* and Carinus. †

The two species of camel, the Bactrian and Arabian, were both well known to the ancients, and are very well described and characterised by Aristotle. ‡

The giraffe, or camelopardalis, was likewise known to the ancients, one having been shewn alive in the circus during the dictatorship of Julius Cæsar, in the year of Rome 708. Ten of them were shewn at once by Gordian III., all of which were slain at the secular games of the emperor Philip. §

When we read with attention the descriptions given of the hippopotamus by Herodotus and Aristotle, which are supposed to have been borrowed from Hecatæus of Miletus, we cannot

---

* See the history of the Hippopotamus, in my second volume.
† Calphurnii, Ecl. VI. 66.
‡ Hist. Anim. lib. II. cap. 1.
§ Jul. Capitol. Gord. III. cap. 23.

fail to perceive that these must have been taken from two very different animals ; one of which is the true hippopotamus, and the other the gnou, or *antilope gnu* of Gmelin's edition of the Systema Naturæ.

The *aper æthiopicus* of Agatharcides, which he describes as having horns, is precisely the Ethiopian hog, or *engallo*, of Buffon and other modern naturalists, whose enormous tusks deserve the name of horns, almost as much as those of the elephant. *

The *bubalus* and the *nagor* are described by Pliny ; the *gazela* by Elian ; the *oryx* by Oppian ; and the *axis*, so early as the time of Ctesias : all of them species of the antelope genus.

Elian gives a very good description of the *bos grunniens*, or grunting ox, under the name of the ox having a tail which serves for a fly-flapper †.

The buffalo was not domesticated by the ancients ; but the *bos Indicus*, or Indian ox of Eli-

* Ælian. Anim. V. 27.

† Id. XV. 14.

an,* having horns sufficiently large to contain three amphoræ, was assuredly that variety of the buffalo which is now called the *arnee*.

The ancients were acquainted with hornless oxen,† and with that African variety of the ox whose horns are only fastened to the skin,‡ and hang down dangling at the sides of the head. They also knew those oxen of India which could run as swift as horses,§ and those which are so small as not to exceed the size of a he-goat.¶ Sheep also with broad tails were not unknown to them,‖ and those other Indian sheep which were as large as asses.**

Although the accounts left us by the ancients respecting the *urus*, or *aurochs*, the rein-deer, and the elk, are all mingled with fable, they are yet sufficient to prove that these animals were not unknown to them, but that the reports which had reached them had been communicated by ignorant or barbarous people, and had not been

* Ælian. Anim. III. 34.
† Id. II. 53.
‡ Id. II. 20.
§ Id. XV. 24.
¶ Id. ibid.
‖ Id. III. 3.
** Id. IV. 32.

corrected by the actual observations of men of learning.

Even the white bear had been seen in Egypt while under the Ptolemies.*

Lions and panthers were quite common at Rome, where they were presented by hundreds in the games of the circus. Even tygers had been seen there, together with the striped hyena, and the nilotic crocodile. There are still preserved in Rome some ancient mosaic, or tesselated pavements, containing excellent delineations of the rarest of these animals; among which a striped hyena is very perfectly represented in a fragment of mosaic, in the Vatican museum. While I was at Rome, a tesselated pavement, composed of natural stones, arranged in the Florentine manner, was discovered in a garden beside the triumphal arch of Galienus, which represented four Bengal tygers in a most admirable manner.

The museum of the Vatican has the figure of

* Athenæis, lib. V.

a crocodile in basalt, almost perfectly represented, except that it has one claw too many on the hind feet. Augustus at one time presented thirty-six of these animals to the view of the people.*

It is hardly to be doubted that the *hippotigris* was the zebra, which is now only found in the southern parts of Africa.† Caracalla killed one of these in the circus.

It might easily be shewn also that almost all the most remarkable species of the *simiæ* of the old world have been distinctly indicated by ancient writers under the names of *pitheci, sphinges, satyri, cephi, cynocephali,* or *cercopitheci.* ‡

They also knew and have described several very small species of *gnawers,* § especially such of that order as possessed any peculiar confor-

---

* Dion. lib. LV.

† Id. LXXVII. Compare also Gisb. Cuperi de Eleph. in nummis obviis. ex. II. cap. 7.

‡ See Lichtenstein, Comment. de Simiarum quotquot veteribus innotuerunt formis. *Hamburg.* 1791.

§ Cuvier gives this name, *rongeurs,* here translated *gnawers* to the order denominated glires by Linnæus, owing to their fore-teeth being peculiarly fitted for gnawing the roots, barks, and stems of vegetables.—*Transl.*

mation or remarkable quality; as we find, for instance, the *jerboa* represented upon the medals of Cyrene, and indicated under the name of *mus bipes*, or two-legged rat. But the smaller species are not of much importance in regard to the object before us, and it is quite sufficient for the enquiry in which we are engaged, to have shewn that all the larger species of quadrupeds, which possess any peculiar or remarkable character, and which we know to inhabit Europe, Asia, and Africa, at the present day, were known to the ancients; whence we may fairly conclude, that their silence in respect to the small quadrupeds, and their neglect in distinguishing the species which very nearly resemble each other, as the various species of antelopes and of some other genera, was occasioned by want of attention and ignorance of methodical arrangement, and not by any difficulties proceeding from the climates or distance of the places which these animals inhabited. We may also conclude with equal certainty, that as eighteen or twenty centuries at the least, with the advantages of circumnavigating Africa, and of penetrating into all the most distant regions of India, have added nothing in this portion of natural history to the information left us by the ancients, it is not at all probable that succeeding ages will add much to the knowledge of our posterity.

Perhaps some persons may be disposed to employ an opposite train of argument, and to allege that the ancients were not only acquainted with as many large quadrupeds as we are, as has been already shewn, but that they actually described several others which we do not now know; that we are rash in considering the accounts of all such animals as fabulous; that we ought to search for them with the utmost care, before concluding that we have acquired a complete knowledge of the existing animal creation; and, in fine, that among these animals which we presume to be fabulous, we may perhaps discover, when better acquainted with them, the actual originals of the bones of those species which are now unknown. Perhaps some may even conceive that the various monsters, essential ornaments of the history of the heroic ages of almost every nation, are precisely those very species which it was necessary to destroy, in order to allow the establishment of civilized societies. Thus Theseus and Bellerophon must have been more fortunate than all the nations of more modern days, who have only been able to drive back the noxious animals into the deserts and ill-peopled regions, but have never yet succeeded in exterminating a single species.

§ 26. *Inquiry respecting the Fabulous Animals of the Ancients.*

It is easy to reply to the foregoing objection, by examining the descriptions that are left us by the ancients of these unknown animals, and by inquiring into their origins. Now the greater number of these animals have an origin purely mythological, and of this origin the descriptions given of them bear the most unequivocal marks; as in almost all of them, we see merely the different parts of known animals united by an unbridled imagination, and in contradiction to every established law of nature.

Those which have been invented by the poetical fancy of the Greeks, have at least some grace and elegance in their composition, resembling the fantastic decorations which are still observable on the ruins of some ancient buildings, and which have been multiplied by the fertile genius of Raphael in his paintings. Like these, they unite forms which please the eye by agreeable contours and fanciful combinations, but which are utterly repugnant to nature and reason; being merely the productions of inventive and playful genius, or perhaps meant as emblematical representations of metaphysical or moral propositions, veiled under mystical hieroglyphics, after the oriental manner. Learned men may be permitted to employ their time and ingenuity in attempts to decypher the mystic knowledge concealed under the forms of the sphynx of Thebes, the pegasus of Thessaly, the minotaur of Crete, or the chimera of Epirus; but it would be folly to expect seriously to find such monsters in nature. We might as well endeavour to find the animals of Daniel, or the beasts of the Apocalypse, in some hitherto unexplored recesses of the globe. Neither can we look for the mythological animals of the Persians,—creatures of a still bolder imagination—such as the *martichore*, or destroyer of men, having a human head on the body of a lion, and the tail of a scorpion; * the *griffin*, or guardian of hidden treasures, half

* Plin. VIII. 21.—Aristot.—Phot. Bibl. art. 72.—Ctes. Indic.—Ælian. Anim. IV. 21.

eagle and half lion; * or the *cartazonon*, or wild ass, armed with a long horn on its forehead. †

Ctesias, who reports these as actual living animals, has been looked upon by some authors as an inventor of fables; whereas he only attributes real existence to hieroglyphical representations. These strange compositions of fancy have been seen in modern times on the ruins of Persepolis. ‡ It is probable that their hidden meanings may never be ascertained; but at all events we are quite certain that they were never intended to be representations of real animals.

Agatharcides, another fabricator of animals, drew his information in all probability from a similar source. The ancient monuments of Egypt still furnish us with numerous fantastic representations, in which the parts of different kinds of creatures are strangely combined—men with the heads of animals, and animals with the heads of

---

* Ælian. Anim.

† Id. XVI. 20.—Photii Bibl. art. 72.—Ctes. Indic.

‡ Le Brun. Voy. to Muscovy, Persia, and India, Vol. II. See also the German work by M. Heeren, on the Commerce of the Ancients.

men; which have given rise to cynocephali, satyrs, and sphinxes. The custom of exhibiting in the same sculpture, in bas-relief, men of very different heights, of making kings and conquerors gigantic, while their subjects and vassals are represented as only a fourth or fifth part of their size, must have given rise to the fable of the pigmies. In some corner of these monuments, Agatharcides must have discovered his carnivorous bull, whose mouth, extending from ear to ear, devoured every other animal that came in his way.* But no naturalist scarcely will acknowledge the existence of any such animal, since nature has never joined cloven hoofs and horns with teeth adapted for cutting and devouring animal food.

There may have been many other figures equally strange with these, either among those monuments of Egypt which have not been able to resist the ravages of time, or in the ancient temples of Ethiopia and Arabia, which have been destroyed by the religious zeal of the Abyssinians and Mahometans. The monuments of India teem

* Phot. Bibl. art. 250.—Agatharcid. Excerp. Hist. cap. 39. —Ælian. Anim. XVII. 45.—Plin. VIII. 21.

with such figures; but the combinations in these are so ridiculously extravagant, that they have never imposed even upon the most credulous. Monsters with an hundred arms, and twenty heads of different kinds, are far too absurd to be believed.

Nay, the inhabitants of China and Japan have their imaginary animals, which they represent as real, and that too in their religious books. The Mexicans had them. In short, they are to be found among every people whose idolatry has not yet acquired some degree of refinement. But is there any one who could possibly pretend to discover, amidst the realities of animal nature, what are thus so plainly the productions of ignorance and superstition? And yet some travellers, influenced by a desire to make themselves famous, have gone so far as to pretend that they saw these fancied beings; or, deceived by a slight resemblance, into which they were too careless to inquire, they have identified these with creatures that actually exist. In their eyes, large baboons, or monkeys, have become *cynocephali*, and sphinxes, real men with long tails. It is thus that St Augustin imagined he had seen a satyr.

Real animals, observed and described with equal inaccuracy, may have given rise to some of these ideal monsters. Thus, we can have no doubt of the existence of the hyena, although the back of this animal be not supported by a single bone, and although it does not change its sex yearly, as alleged by Pliny. Perhaps the carnivorous bull may only have been the two-horned rhinoceros, falsely described. M. de Weltheim considers the auriferous ants of Herodotus as the *corsacs* * of modern naturalists.

The most famous among these fabulous animals of the ancients was the *unicorn*. Its real existence has been obstinately asserted even in the present day, or at least proofs of its existence have been eagerly sought for. Three several animals are frequently mentioned by the ancients as having only one horn placed on the middle of the forehead. The *oryx* of Africa, having cloven hoofs, the hair placed reversely to that of other animals, † its height equal to that of the

* The Korsake, or Corsac fox of Pallas and Pennant.—*Transl.*

† Aristot. Anim. II. 1. and III. 2.—Plin. XI. 46.

bull, * or even of the rhinoceros, † and said to resemble deer and goats in its form; ‡ the *Indian ass*, having solid hoofs; and the *monoceros*, properly so called, whose feet are sometimes compared to those of the lion, § and sometimes to those of the elephant, || and is therefore considered as having divided feet. The horse unicorn ¶ and the bull-unicorn are doubtless both referable to the Indian ass, for even the latter is described as having solid hoofs. ** We may therefore be fully assured that these animals have never really existed, as no solitary horns have ever found their way into our collections, excepting those of the rhinoceros and narwal.

After careful consideration, it is impossible that we should give any credit to rude sketches made by savages upon rocks. Entirely ignorant of perspective, and wishing to represent the outlines of a straight-horned antelope in profile, they could only give the figure one horn, and thus

---

* Herodot. IV. 192.
† Oppian, Cyneg. II. vers. 551.
‡ Plin. VIII. 53.
§ Philostrog. III. ii.
|| Plin. VIII. 21.
¶ Onesecrit. ap. Strab. lib. XV.—Ælian. Anim. XIII. 42.
** See Pliny and Solinus.

they produced an oryx. The oryxes, too, that are seen on the Egyptian monuments, are nothing more, probably, than productions of the stiff style, imposed on the sculptors of the country, by religious prejudices. Several of their profiles of quadrupeds shew only one fore and one hinder leg, and it is probable that the same rule led them also to represent only one horn. Perhaps their figures may have been copied after individuals that had lost one of their horns by accident, a circumstance that often happens to the chamois and the saiga, species of the antelope genus, and this would be quite sufficient to establish the error. All the ancients, however, have not represented the oryx as having only one horn. Oppian expressly attributes two to this animal, and Ælian mentions one that had four. * Finally, if this animal was ruminant and cloven-footed, we are quite certain that its frontal bone must have been divided longitudinally into two, and that it could not possibly, as is very justly remarked by Camper, have had a horn placed upon the suture.

* Ælian. Anim. XV. 14.

It may be asked, however, What two horned animal could have given an idea of the *oryx*, in the forms in which it has been transmitted down to us, even independent of the notion of a single horn? To this I answer, as already done by Pallas, that it was the straight-horned *antilope oryx* of Gmelin, improperly named *pasan* by Buffon. This animal inhabits the deserts of Africa, and must frequently approach the confines of Egypt, and appears to be that which is represented in the hieroglyphics. It equals the ox in height, while the shape of its body approaches to that of a stag, and its straight horns present exceedingly formidable weapons, hard almost as iron and sharp-pointed like javelins. Its hair is whitish; it has black spots and streaks on its face, and the hair on its back points forwards. Such is the description given by naturalists; and the fables of the Egyptian priests, which have occasioned the insertion of its figure among their hieroglyphics, do not require to have been founded in nature. Supposing that an individual of this species may have been seen which had lost one of its horns by some accident, it may have been taken as a representative of the entire race, and erroneously adopted by Aristotle to be copied by all his successors. All this is quite possible and even natural, and gives not the smallest evidence for the existence of a single-horned species of antelope.

In regard to the Indian ass, of the alexipharmic virtues of whose horn the ancients speak, we find the eastern nations of the present day attributing exactly the same properties of counteracting poison to the horn of the rhinoceros. When this horn was first imported into Greece, nothing probably was known respecting the animal to which it belonged; and accordingly it was not known to Aristotle. Agatharcides is the first author by whom it is mentioned. In the same manner, ivory was known to the ancients long before the animal from which it is procured; and perhaps some of their travellers may have given to the rhinoceros the name of *Indian ass*, with as much propriety as the Romans denominated the elephant the *bull of Lucania*. Every thing which they relate of the strength, size, and ferocity of their wild ass of India, corresponds sufficiently with the rhinoceros. In succeeding times, when the rhinoceros came to be better known to naturalists, finding that former authors mentioned a single-horned animal under the name of Indian ass, they concluded, without any examination, that it must be quite a distinct creature, having solid hoofs. We have remaining a detailed description of the Indian ass, written by Ctesias; * but, as we have already seen that this must have been taken from the ruins of Persepolis, it should go for nothing in the real history of the animal.

When there afterwards appeared more exact descriptions of an animal having several toes or hoofs on each foot, the ancients conceived it to be a third species of one-horned animals, to which they gave the name of *monoceros*. These double, and even triple references, are more frequent among ancient writers, because most of their works which have come down to us were mere compilations; because even Aristotle himself has often mixed borrowed facts with those which had come under his own observation; and because the habit of critically investigating the authorities of previous writers, was as little known among ancient naturalists as among their historians.

From all these reasonings and digressions, it may be fairly concluded, that the large animals

---

* Ælian. Anim. IV. 32.

of the ancient continent with which we are now acquainted, were known to the ancients; and that all the animals of which the ancients have left descriptions, and which are now unknown, were merely fabulous. It also follows, that the large animals of the three anciently known quarters of the world, were very soon known to the people who frequented their coasts.

It may also be concluded, that no large species remain to be discovered in America, as there is no good reason that can be assigned why any such should exist in that country with which we are unacquainted, and in fact none has been discovered there during the last hundred and fifty years. The tapir, jaguar, puma, cabiai or capibara, glama, vicunna, red-wolf, buffalo, or American bison, ant-eaters, sloths, and armadillos, are all contained in the works of Margrave and Hernandez, as well described as in Buffon and even better, for Buffon has mistaken and confused the natural history of the ant-eaters, has mixed the description of the jaguar with that of the red-wolf, and has confounded the American bison with the aurochs, or urus, of Poland. Pennant, it is true, was the first naturalist who clearly distinguished the musk ox, but it had been long mentioned by travellers. The cloven-footed, or Chilese, horse of Molina, has not been described by any of the early Spanish travellers, but its existence is more than doubtful, and the authority of Molina is too suspicious to entitle us to believe that this animal actually exists. The Muflon of the blue mountains is the only American quadruped of any size hitherto known, of which the discovery is entirely modern; and perhaps it may only have been an *argali*, that had strayed from eastern Siberia over the ice. *

After all that has been said, it is quite impossible to conceive that the enormous *mastedontes* and gigantic *megatheria*, † whose bones have been discovered under ground in North and South America, can still exist alive in that quarter of the world. They could not fail to be observed by the hunting tribes, which continually wander

* The argali had long before been mentioned by writers as inhabiting Kamtschatka, the Kurili islands, and probably the north-west coast of America and California.—*Transl.*

† These are new names devised to characterise the animals of which the bones and teeth have been found in large quantities in America, both in Virginia on the banks of the Ohio, and in Chili and Peru.—*Transl.*

in all directions through the wilds of America. Indeed they themselves seem to be fully aware that these animals no longer exist in their country, as they have invented a fabulous account of their destruction, alleging that they were all killed by the Great Spirit, to prevent them from extirpating the human race. It is quite obvious, that this fable has been invented subsequently to the discovery of the bones; just as the inhabitants of Siberia have contrived one respecting the *mammoth*, whose bones have been found in that country, alleging that it still lives under ground like the mole: and just as the ancients had their fables about the graves of giants, who were thought to have been buried wherever the bones of elephants happened to be dug up.

From all these considerations, it may be safely concluded, as shall be more minutely explained in the sequel,—That none of the large species of quadrupeds, whose remains are now found imbedded in regular rocky strata, are at all similar to any of the known living species:—That this circumstance is by no means the mere effect of chance, or because the species to which these fossil bones have belonged are still concealed in the desert and uninhabited parts of the world, and have hitherto escaped the observation of travellers; but,—That this astonishing phenomenon has proceeded from general causes, and that the careful investigation of it affords one of the best means for discovering and explaining the nature of these causes.

§ 27. *Of the Difficulty of distinguishing the Fossil Bones of Quadrupeds.*

While the study of the fossil remains of the greater quadrupeds is more satisfactory, by the clear results which it affords, than that of the remains of other animals found in a fossil state, it is also complicated with greater and more numerous difficulties. Fossil shells are usually found quite entire, and retaining all the characters requisite for comparing them with the specimens contained in collections of natural history, or represented in the works of naturalists. Even the skeletons of fishes are found more or less entire, so that the general forms of their bodies can, for the most part, be ascertained, and usually at least their generic and specific characters are determinable, as these are mostly drawn from their solid parts. In quadrupeds, on the contrary, even when their entire skeletons are found, there is great difficulty in discovering their distinguishing characters, as these are chiefly founded upon their hair and colours, and other marks which have disappeared previous to their incrustation. It is also very rare to find any fossil skeletons of quadrupeds in any degree approaching to a complete state, as the strata for the most part only contain separate bones, scattered confusedly, and almost always broken and reduced to fragments, which are the only means left to naturalists for ascertaining the species or genera to which they have belonged.

It may be stated also, that most observers, alarmed by these formidable difficulties, have passed slightly over the fossil remains of quadrupeds, and have satisfied themselves with classing them vaguely, by means of slight resemblances, or have not even pretended to give them names. Hence this portion of the history of extraneous fossils, though the most important and most instructive, has been investigated with less care than any other. *

---

* As I have already remarked on a former occasion, it is not my intention, by these observations, to detract from the merits

Fortunately, comparative anatomy, when thoroughly understood, enables us to surmount all these difficulties, as a careful application of its principles instructs us in the correspondence and dissimilarity of the forms of organized bodies of different kinds, by which each may be rigorously ascertained, from almost every fragment of its various parts and organs.

Every organized individual forms an entire system of its own, all the parts of which mutually correspond, and concur to produce a certain definite purpose, by reciprocal reaction, or by combining towards the same end. Hence none of these separate parts can change their forms without a corresponding change on the other parts of the same animal, and consequently each of these parts taken separately, indicates all the other parts to which it has belonged. Thus, as I have elsewhere shewn, if the viscera of an animal are so organized as only to be fitted for

---

of Camper, Pallas, Blumenbach, Sœmmering, Merk, Faujas, Rosenmuller, and other naturalists, in regard to extraneous fossils: But, though their observations have been of great value in my researches, and are quoted by me in every step, they are in general very incomplete.

the digestion of recent flesh, it is also requisite that the jaws should be so constructed as to fit them for devouring prey; the claws must be constructed for seizing and tearing it to pieces; the teeth for cutting and dividing its flesh; the entire system of the limbs, or organs of motion, for pursuing and overtaking it; and the organs of sense, for discovering it at a distance. Nature also must have endowed the brain of the animal with instincts sufficient for concealing itself, and for laying plans to catch its necessary victims.

Such are the universal conditions that are indispensable in the structure of carnivorous animals; and every individual of that description must necessarily possess them combined together, as the species could not otherwise subsist. Under this general rule, however, there are several particular modifications, depending upon the size, the manners, and the haunts of the prey for which each species of carnivorous animal is destined or fitted by nature; and, from each of these particular modifications, there result certain differences in the more minute conformations of particular parts, all, however, comformable to the general principles of structure already mentioned. Hence it follows, that in every one of their parts we discover distinct indications, not only of the classes and orders of animals, but also of their genera, and even of their species.

In fact, in order that the jaw may be well adapted for laying hold of objects, it is necessary that its condyle should have a certain form; that the resistance, the moving power, and the fulcrum, should have a certain relative position with respect to each other; and that the temporal muscles should be of a certain size: The hollow or depression, too, in which these muscles are lodged, must have a certain depth; and the zygomatic arch under which they pass, must not only have a certain degree of convexity, but it must be sufficiently strong to support the action of the masseter.

To enable the animal to carry off its prey when seized, a correspondent force is requisite in the muscles which elevate the head; and this necessarily gives rise to a determinate form of the vertebræ to which these muscles are attached, and of the occiput into which they are inserted.

In order that the teeth of a carnivorous animal may be able to cut the flesh, they require to be sharp, more or less so in proportion to the greater or less quantity of flesh that they have to cut. It is requisite that their roots shonld be solid and strong, in proportion to the quantity and the size of the bones which they have to break in pieces. The whole of these circumstances must necessarily influence the developement and form of all the parts which contribute to move the jaws.

To enable the claws of a carnivorous animal to seize its prey, a considerable degree of mobility is necessary in their paws and toes, and a considerable strength in the claws themselves. From these circumstances, there necessarily result certain determinate forms in all the bones of their paws, and in the distribution of the muscles and tendons by which they are moved. The fore-arm must possess a certain facility of moving in various directions, and consequently requires certain determinate forms in the bones of which it is composed. As the bones of the fore-arm are articulated with the arm-bone or humerus, no change can take place in the form and structure of the former, without occasioning correspondent changes in the form of the latter. The shoulder blade also, or scapula, requires a correspondent degree of strength in all animals destined for catching prey, by which it likewise must necessarily have an appropriate form. The play and action of all these parts require certain proportions in the muscles which set them in motion, and the impressions formed by these muscles must still farther determine the forms of all these bones.

After these observations, it will be easily seen that similar conclusions may be drawn with respect to the hinder limbs of carnivorous animals, which require particular conformations to fit them for rapidity of motion in general; and that similar considerations must influence the forms and connections of the vertebræ and other bones constituting the trunk of the body, to fit them for flexibility and readiness of motion in all directions. The bones also of the nose, of the orbit, and of the ears, require certain forms and structures to fit them for giving perfection to the senses of smell, sight, and hearing, so necessary to animals of prey. In short, the shape and structure of the teeth regulate the forms of the condyle, of the shoulder-blade, and of the claws, in the same manner as the equation of a curve regulates all its other properties; and, as in regard to any particular curve, all its properties may be ascertained by assuming each separate property as the foundation of a particular equation; in the same manner, a claw, a shoulder-blade, a condyle, a leg or arm bone, or any other bone, separately considered, enables us to discover the description of teeth to which they have belonged; and so also reciprocally we may determine the forms of the other bones from the teeth. Thus, commencing our investigation by a careful survey of any one bone by itself, a person who is sufficiently master of the laws of organic structure, may, as it were, reconstruct the whole animal to which that bone had belonged.

This principle is sufficiently evident, in its general acceptation, not to require any more minute demonstration; but, when it comes to be applied in practice, there is a great number of cases in which our theoretical knowledge of these relations of forms is not sufficient to guide us, unless assisted by observation and experience.

For example, we are well aware that all hoofed animals must necessarily be herbivorous, because they are possessed of no means of seizing upon prey. It is also evident, having no other use for their fore-legs than to support their bodies, that they have no occasion for a shoulder so vigorously organized as that of carnivorous animals; owing to which, they have no clavicles or accromion processes, and their shoulder-blades are proportionally narrow. Having also no occasion to turn their fore-arms, their radius is joined by ossification to the ulna, or is at least articulated by gynglymus with the humerus. Their food being entirely herbaceous, requires teeth with flat surfaces, on purpose to bruise the seeds and plants on which they feed. For this purpose also, these surfaces require to be unequal, and are consequently composed of alternate perpendicular layers of hard enamel and softer bone. Teeth of this structure necessarily require horizontal motions, to enable them to triturate or grind down the herbaceous food; and, accordingly, the condyles of the jaw could not be formed into such confined joints as in the carnivorous animals, but must have a flattened form, correspondent to sockets in the temporal bones, which also are more or less flat for their reception. The hollows likewise of the temporal bones, having smaller muscles to contain, are narrower, and not so deep, &c. All these circumstances are deducible from each other, according to their greater or less generality, and in such manner that some are essentially and exclusively appropriated to hoofed quadrupeds, while other circumstances, though equally necessary to that description of animals, are not exclusively so, but may be found in animals of other descriptions, where other conditions permit or require their existence.

When we proceed to consider the different orders or subdivisions of the class of hoofed animals, and examine the modifications to which the general conditions are liable, or rather the particular conditions which are conjoined, according to the respective characters of the several subdivisions, the reasons upon which these particular conditions or rules of conformation are founded become less evident. We can easily conceive, in general, the necessity of a more complicated system of digestive organs in those species which have less perfect masticatory systems; and hence we may presume that these latter animals require especially to be ruminant, which are in want of such or such kinds of teeth; and may also deduce, from the same considerations, the necessity of a certain conformation of the esophagus, and of corresponding forms in the vertebræ of the neck, &c. But I doubt whether it would have been discovered, independently of actual observation, that ruminant animals should all have cloven hoofs, and that they should be the only animals having that particular conformation; that the ruminant animals only should be provided with horns on their foreheads; that those among them which have sharp tusks, or canine teeth, should want horns, &c.

As all these relative conformations are constant and regular, we may be assured that they depend upon some sufficient cause; and, since we are not acquainted with that cause, we must here supply the defect of theory by observation, and in this way lay down empirical rules on the subject, which are almost as certain as those deduced from rational principles, especially if established upon careful and repeated observation. Hence, any one who observes merely the print of a cloven hoof, may conclude that it has been left by a ruminant animal, and regard the conclusion as equally certain with any other in physics or in morals. Consequently, this single foot-mark clearly indicates to the observer the forms of the teeth, of the jaws, of the vertebræ, of all the leg-bones, thighs, shoulders, and of the trunk of the body of the animal which left the mark. It is much surer than all the marks of Zadig. Observation alone, independent entirely of general principles of philosophy, is sufficient to shew that there certainly are secret reasons for all these relations of which I have been speaking.

When we have established a general system of these relative conformations of animals, we not only discover specific constancy, if the expression may be allowed, between certain forms of certain organs, and certain other forms of different organs; we can also perceive a classified constancy of conformation, and a correspondent gradation between these two sets of organs, which demonstrate their mutual influence upon each other, almost as certainly as the most perfect deduction of reason. For example, the masticatory system is generally more perfect in the non-ruminant hoofed quadrupeds than it is in the cloven-hoofed or ruminant quadrupeds; as the former possess incisive teeth, or tusks, or almost always both of these, in both jaws. The structure also of their feet is in general more complicated, having a greater number of toes, or their phalanges less enveloped in the hoof, or a greater number of distinct metacarpal and metatarsal bones, or more numerous tarsal bones, or the fibula more completely distinct from the tibia: or, finally, that all these enumerated circumstances are often united in the same species of animal.

It is quite impossible to assign reasons for these relations; but we are certain that they are not produced by mere chance, because, whenever a cloven-hoofed animal has any resemblance in the arrangement of its teeth to the animals we now speak of, it has the resemblance to them also in the arrangement of its feet. Thus camels, which have tusks, and also two or four incisive teeth in the upper jaw, have one additional bone in the tarsus, their scaphoid and cuboid bones not being united into one; and have also very small hoofs, with corresponding phalanges, or toe-bones. The musk animals, whose tusks are remarkably conspicuous, have a distinct fibula as long as the tibia; while the other cloven-footed animals have only a small bone articulated at the lower end of the tibia, in place of a fibula. We have thus a constant mutual relation between the organs or conformations, which appear to have no kind of connection with each other; and the gradations of their forms invariably correspond, even in those cases in which we cannot give the rationale of their relations.

By thus employing the method of observation, where theory is no longer able to direct our views, we procure astonishing results. The smallest fragment of bone, even the most apparently insignificant apophysis, possesses a fixed and determinate character, relative to the class, order, genus, and species of the animal to which it belonged; insomuch, that, when we find merely the extremity of a well-preserved bone, we are able, by careful examination, assisted by analogy and exact comparison, to determine the species to which it once belonged, as certainly as if we had the entire animal before us. Before venturing to put entire confidence in this method of investigation, in regard to fossil bones, I have very frequently tried it with portions of bones belonging to well-known animals, and always with such complete success, that I now entertain no doubt with regard to the results which it affords. I must acknowledge that I enjoy every kind of advantage for such investigations that could possibly be of use, by my fortunate situation in the Museum of Natural History; and, by assiduous researches for nearly fifteen years, I have collected skeletons of all the genera and sub-genera of quadrupeds, with those of many species in some of the genera and even of several varieties of some species. With these aids, I have found it easy to multiply comparisons, and to verify, in every point of view, the application of the foregoing rules.

We cannot, in the present Essay, enter into a more lengthened detail of this method, and must refer for its entire explanation to the large work on Comparative Anatomy, which we propose to publish very soon, and in which all its laws will be explained and illustrated. In the meantime, the intelligent reader may gather a great number of these from the work now laid before him, if he will take the trouble of attending to all the applications which we have made of them. He will there find that it is by this method alone that we have been guided, and that it has almost always been sufficient for the purpose of referring every fossil bone to its peculiar species, if belonging to one that still exists; to its genus, if belonging to an unknown species; to its order, if belonging to a new genus; and, finally, to its class, if belonging to an unknown order: And, in these three latter predicaments, to assign to it the proper characters for distinguishing it from the nearest resembling orders, genera, and species. Before the commencement of these researches, naturalists had done no more than this in regard even to such animal remains as were found in an entire state.

§ 28. *Results of the Researches respecting the Fossil Bones of Quadrupeds.*

In this manner we have ascertained and classified the fossil remains of seventy-eight different quadrupeds, in the viviparous and oviparous classes. Of these, forty-nine are distinct species hitherto entirely unknown to naturalists. Eleven or twelve others have such entire resemblance to species already known, as to leave no doubts whatever of their identity; and the remaining sixteen or eighteen have considerable traits of resemblance to known species, but the comparison of these has not yet been made with so much precision as to remove all dubiety.

Of the forty-nine new or hitherto unknown

species, twenty-seven are necessarily referable to seven new genera; while the other twenty-two new species belong to sixteen genera, or sub-genera, already known. The whole number of genera and sub-genera to which the fossil remains of quadrupeds hitherto investigated are referable, are thirty-six, including those belonging both to known and unknown species.

Of these seventy-eight species, fifteen which belong to eleven genera or sub-genera, are animals belonging to the class of oviparous quadrupeds; while the remaining sixty-three belong to the mammiferous class. Of these last, thirty-two species are hoofed animals, not ruminant, and reducible to ten genera; twelve are ruminant animals, belonging to two genera; seven are *gnawers*, referable to six genera; eight are carnivorous quadrupeds, belonging to five genera; two are toothless animals, of the sloth genus; and two are amphibious animals of two distinct genera. *

* As the author has already referred fifteen other species to what he terms the oviparous class of quadrupeds, the two *amphibious* animals here mentioned probably belong to the or-

§ 29. *Relations of the Species of Fossil Bones, with the Strata in which they are found.*

Notwithstanding the considerable number of these fossil bones already discovered and ascertained, it would be premature to attempt establishing any conclusions deduced from them in regard to the theory of the earth, as they are not in sufficient proportion to the entire number of genera and species which, in all probability, are buried in the strata of the earth. Hitherto the bones of the larger species have chiefly been collected, as more obvious to the labourers, while those of smaller animals are usually neglected, unless when they fall by accident in the way of a naturalist, or when some other remarkable circumstance, such as their extreme abundance in any particular place, attracts even the attention of common people.

The most important consideration, that which

der of cetaceous mammiferous animals, and not to the *amphibiæ* of the Linnæan system.—*Transl.*

has been the chief object of my researches, and which constitutes their legitimate connection with the theory of the earth, is to ascertain the particular strata in which each of the species was found, and to inquire if any of the general laws could be ascertained, relative either to the zoological subdivisions, or to the greater or less resemblance between these fossil species and those which still exist upon the earth.

The laws already recognised with respect to these relations are very distinct and satisfactory.

It is, in the first place, clearly ascertained, that the oviparous quadrupeds are found considerably earlier, or in more ancient strata, than those of the viviparous class. Thus the crocodiles of Honfleur and of England are found underneath the chalk. The *monitors** of Thuringia would be still more ancient, if, according to the Wernerian school, the copper-slate in which they are contained, along with a great number of fishes supposed to have belonged to fresh water, is to be placed among the most ancient strata of the

* The lacerta monitor of Linnæus.—*Transl.*

secondary formations. The great alligators, or crocodiles, and the tortoises of Maestricht, are found in the chalk formations; but these are both marine animals.

This earliest appearance of fossil bones seems to indicate, that dry lands and fresh waters must have existed before the formation of the chalk strata. Yet neither at that early epoch, nor during the formation of the chalk strata, nor even for a long period afterwards, do we find any fossil remains of mammiferous land-quadrupeds.

We begin to find the bones of mammiferous sea-animals, namely, of the lamantin and of seals, in the coarse shell limestone which immediately covers the chalk strata in the neighbourhood of Paris. But no bones of mammiferous land-quadrupeds are to be found in that formation; and, notwithstanding the most careful investigations, I have never been able to discover the slightest traces of this class, except in the formations which lie over the coarse limestone strata; but immediately on reaching these more recent formations, the bones of land-quadrupeds are discovered in great abundance.

As it is reasonable to believe that shells and

fish did not exist at the period of the formation of the primitive rocks, we are also led to conclude that the oviparous quadrupeds began to exist along with the fishes, and at the commencement of the period which produced the secondary formations; while the land quadrupeds did not appear upon the earth till long afterwards, and until the coarse shell limestone had been already deposited, which contains the greater part of our genera of shells, although of quite different species from those that are now found in a natural state.

It is remarkable that those coarse limestone strata, which are chiefly employed at Paris for building, are the last-formed strata which indicate a long and quiet continuance of the water of the sea above the surface of our continent. Above them, indeed, there are found formations containing abundance of shells and other productions of the sea; but these consist of alluvial materials, sand, marle, sandstone, or clay, which rather indicate transportations that have taken place with some degree of violence, than strata formed by quiet depositions; and where some regular rocky strata, of inconsiderable extent and thickness, appear above or below these alluvial formations, they generally bear the marks of having been deposited from fresh water.

All the known specimens of the bones of viviparous land-quadrupeds, have either been found in these formations from fresh water, or in the alluvial formations; whence there is every reason to conclude that these animals have only begun to exist, or at least to leave their remains in the strata of our earth, since the last retreat of the sea but one, and during that state of the world which preceded its last irruption.

There is also a determinate order observable in the disposition of these bones in regard to each other, which indicates a very remarkable succession in the appearance of the different species. All the genera which are now unknown, as the *palæotheria, anaplotheria*, &c. with the localities of which we are thoroughly acquainted, are found in the most ancient of those formations of which we are now treating, or those which are placed directly over the coarse limestone strata. It is chiefly they which occupy the regular strata that have been deposited from fresh water, or certain alluvial beds of very ancient formation, generally composed of sand

and rounded pebbles; which were perhaps the earliest alluvial formations of the ancient world. Along with these there are also found some lost species of known genera, but in small numbers; together with some oviparous quadrupeds and some fish, which appear to have been inhabitants of fresh water. The strata containing these are always more or less covered with alluvial formations, filled with shells and other productions of the sea.

The most celebrated of the unknown species belonging to known genera, or to genera nearly allied to those that are known, as the fossil elephant, rhinoceros, hippopotamus, and *mastodon*, are never found along with the more ancient genera; but are only contained in alluvial formations, sometimes along with sea shells, and sometimes with fresh-water shells, but never in regular rocky strata. Every thing found along with these species is either, like them, unknown, or at least doubtful.

Lastly, the bones of species which are apparently the same with those that still exist alive, are never found except in the very latest alluvial depositions, or those which are either formed on the sides of rivers, or on the bottoms of ancient lakes or marshes now dried up, or in the substance of beds of peat, or in the fissures and caverns of certain rocks, or at small depths below the present surface, in places where they may have been overwhelmed by debris, or even buried by man: And, although these bones are the most recent of all, they are almost always, owing to their superficial situation, the worst preserved.

It must not, however, be thought that this classification of the various mineral repositories is as certain as that of the species, and that it has nearly the same character of demonstration. Many reasons might be assigned to shew that this could not be the case. All the determinations of species have been made, either by means of the bones themselves, or from good figures; whereas it has been impossible for me personally to examine the places in which these bones were found. Indeed I have often been reduced to the necessity of satisfying myself with vague and ambiguous accounts, given by persons who did not know well what was necessary to be noticed; and I have still more frequently been unable to procure any information whatever on the subject.

Secondly, these mineral repositories are subject to infinitely greater doubts in regard to their successive formations, than are the fossil bones respecting their arrangement and determination. The same formation may seem recent in those places where it happens to be superficial, and ancient where it has been covered over by succeeding formations. Ancient formations may have been transported into new situations by means of partial inundations, and may thus have covered over recent formations containing bones; they may have them carried over them by debris, so as to surround these recent bones, and may have mixed with them the productions of the ancient sea, which they previously contained. Anciently-deposited bones may have been washed out from their original situations by the waters, and been afterwards, enveloped in recent alluvial formations. And, lastly, recent bones may have fallen into the crevices and caverns of ancient rocks, where they may have been covered up by stalactites or other incrustations. In every individual instance, therefore, it becomes necessary to examine and appreciate all these circumstances, which might otherwise conceal the real origin of extraneous fossils; and it rarely happens that the people who found these fossil bones were aware of this necessity, and consequently the true characters of their repositories have almost always been overlooked or misunderstood.

Thirdly, there are still some doubtful species of these fossil bones, which must occasion more or less uncertainty in the result of our researches, until they have been clearly ascertained. Thus the fossil bones of horses and buffaloes, which have been found along with those of elephants, have not hitherto presented sufficiently distinct specific characters; and such geologists as are disinclined to adopt the successive epochs which I have endeavoured to establish in regard to fossil bones, may for many years draw from thence an argument against my system, so much the more convenient as it is contained in my own work. Even allowing that these epochs are liable to some objections, from such as have slightly considered some particular fact, I am not the less satisfied that those who shall take a comprehensive view of the phenomena, will not be checked by inconsiderable and partial difficulties, but will be led to conclude, as I have done, that there has at least been one succession, and very probably two, in the class of quadrupeds, before the appearance of those races which now inhabit the surface of our globe.

§ 30. *Proofs that the extinct Species of Quadrupeds are not Varieties of the presently existing Species.*

The following objection has already been started against my conclusions. Why may not the presently existing races of mammiferous land quadrupeds be mere modifications or varieties of those ancient races which we now find in the fossil state, which modifications may have been produced by change of climate and other local circumstances, and since raised to the present excessive difference, by the operation of similar causes during a long succession of ages?

This objection may appear strong to those who believe in the indefinite possibility of change of forms in organized bodies, and think that during a succession of ages, and by alterations of habitudes, all the species may change into each other, or one of them give birth to all the rest. Yet to these persons the following answer may be given from their own system: If the species have changed by degrees, as they assume, we ought to find traces of this gradual modification. Thus, between the *palæotherium* and the species of our own days, we should be able to discover some intermediate forms; and yet no such discovery has ever been made. Since the bowels of the earth have not preserved monuments of this strange genealogy, we have a right to conclude, That the ancient and now extinct species were as permanent in their forms and characters as those which exist at present; or at least, That the catastrophe which destroyed them did not leave sufficient time for the production of the changes that are alleged to have taken place.

In order to reply to those naturalists who acknowledge that the varieties of animals are restrained by nature within certain limits, it would be necessary to examine how far these limits extend. This is a very curious inquiry, and in itself exceedingly interesting under a variety of relations, but has been hitherto very little attended to. It requires that we should define accurately what is, or ought to be, understood by the word species, which may be thus expressed :— *A species comprehends all the individuals which descend from each other, or from a common parentage, and those which resemble them as much as they do each other.* Thus the different races which have been generated from them, are considered as varieties but of one species. Our observations, therefore, respecting the differences between the ancestors and the descendants, are the only rules by which we can judge on this subject; all other considerations being merely hypothetical, and destitute of proof. Taking the word *variety* in this limited sense, we observe that the differences which constitute this variety depend upon determinate circumstances, and that their extent increases in proportion to the intensity of the circumstances which occasion them.

Upon these principles it may be observed, that the most superficial characters are the most variable. Thus colour depends much upon light; thickness of hair upon heat; size upon abundance of food, &c. In wild animals, however, even these varieties are greatly limited by the natural habits of the animal, which does not willingly migrate from the places where it finds in sufficient quantity what is necessary for the support of its species, and does not even extend its haunts to any great distances, unless it also finds all these circumstances conjoined. Thus, although the wolf and the fox inhabit all the climates from the torrid to the frigid zone, we hardly find any other differences among them, through the whole of that vast space, than a little more or a little less beauty in their furs. I have compared the skulls of foxes from the most northern regions and from Egypt, with those of France, and found no differences but what might naturally be expected in different individuals. The more savage animals, especially those which are carnivorous, being confined within narrower limits, vary still less; and the only difference between the hyena of Persia and that of Morocco, consists in a thicker or a thinner mane.

Wild animals which subsist upon herbage feel the influence of climate a little more extensively, because there is added to it the influence of food, both in regard to its abundance and its quality. Thus the elephants of one forest are larger than those of another; their tusks also grow somewhat longer in places where their food may happen to be more favourable for the production of the substance of ivory. The same may take place in regard to the horns of stags and rein-deer. But let us examine two elephants the most dissimilar that can be conceived, we shall not discover the smallest difference in the number and articulations of the bones, the structure of the teeth, &c.

Besides, the species of herbivorous animals, in their wild state, seem more restrained from migrating and dispersing than the carnivorous species, being influenced both by climate and by the kind of nourishment which they need.

Nature appears also to have guarded against the alterations of species which might proceed from mixture of breeds, by influencing the various species of animals with mutual aversion from each other. Hence all the cunning and all the force that man is able to exert is necessary to accomplish such unions, even between species that have the nearest resemblances. And when the mule-breeds that are thus produced by these forced conjunctions happen to be fruitful, which is seldom the case, this fecundity never continues beyond a few generations, and would not probably proceed so far, without a continuance of the same cares which excited it at first. Thus we never see in a wild state intermediate productions between the hare and the rabbit, between the stag and the doe, or between the martin and the weasel. But the power of man changes this established order, and contrives to produce all these intermixtures of which the various species are susceptible, but which they would never produce if left to themselves.

The degrees of these variations are proportional to the intensity of the causes that produce them, namely, the slavery or subjection under which those animals are to man. They do not proceed far in half-domesticated species. In the cat, for example, a softer or harsher fur, more brilliant or more varied colours, greater or less size—these form the whole extent of the varieties in the species; the skeleton of the cat of Angora differs in no regular and constant circumstances from the wild cat of Europe.

In the domesticated herbivorous quadrupeds, which man transports into all kinds of climates, and subjects to various kinds of management, both in regard to labour and nourishment, he procures certainly more considerable variations, but still they are all merely superficial. Greater or less size; longer or shorter horns, or even the want of these entirely; a hump of fat, larger or smaller, on the shoulder; these form the chief differences among particular races of the *bos taurus*, or domestic black cattle; and these differences continue long in such breeds as have been transported to great distances from the countries in which they were originally produced, when proper care is taken to prevent crossing.

The innumerable varieties in the breeds of the *ovis aries*, or common sheep, are of a similar nature, and chiefly consist in differences of their fleeces, as the wool which they produce is a very important object of attention. These varieties, though not quite so perceptible, are yet sufficiently marked among horses. In general the forms of the bones are very little changed; their connections and articulations, and the form and structure of the large grinding teeth, are invariably the same. The small size of the tusks in the domesticated hog, compared with the wild boar, of which it is only a cultivated variety, and the junction of its cloven hoofs into one solid hoof, observable in some races, form the extreme point of the differences which man has been able to produce among herbivorous domesticated quadrupeds.

The most remarkable effects of the influence of man is produced upon that animal which he has reduced most completely under subjection. Dogs have been transported by mankind into every part of the world, and have submitted their actions to his entire direction. Regulated in their sexual unions by the pleasure or caprice of their masters, the almost endless varieties of dogs differ from each other in colour; in length and abundance of hair, which is sometimes entirely wanting; in their natural instincts; in size, which varies in measure as one to five, amounting, in some instances, to more than an hundred fold in bulk; in the forms of their ears, noses, and tails; in the relative length of their legs; in the progressive developement of the brain in several of the domesticated varieties, occasioning alterations even in the form of the head; some of them having long slender muzzles with a flat forehead; others having short muzzles, with the forehead convex, &c. insomuch that the apparent differences between a mastiff and a water-spaniel, and between a greyhound and a pug-dog, are even more striking than between almost any of the wild species of a genus. Finally, and this may be considered as the maximum of known variation in the animal kingdom, some races of dogs have an additional claw on each hind-foot, with corresponding bones of the tarsus; as there sometimes occur in the human species some families that have six fingers on each hand. Yet, in all these varieties, the relations of the bones with each other remain essentially the same, and the form of the teeth never changes in any perceptible degree, except that in some individuals one additional false grinder occasionally appears, sometimes on the one side and sometimes on the other. *

It follows from these observations, that animals have certain fixed and natural characters, which resist the effects of every kind of influence, whether proceeding from natural causes or human interference; and we have not the smallest reason to suspect that time has any more effect upon them than climate.

---

* See, in the Annals of the Museum, XVIII. 333., a memoir by my brother on the varieties of dogs, which he drew up at my request, from a series of skeletons of all the varieties of dogs, prepared by me expressly on purpose.

I am well aware that some naturalists lay prodigious stress on the thousands of years which they can call into action by a dash of their pens. In such matters, however, our only way of judging as to the effects which may be produced by a long period of time, is by multiplying, as it were, such as are produced by a shorter known time. With this view I have endeavoured to collect all the ancient documents respecting the forms of animals; and there are none equal to those furnished by the Egyptians, both in regard to their antiquity and abundance. They have not only left us representations of animals, but even their identical bodies embalmed and preserved in the catacombs.

I have examined with the greatest attention the engraved figures of quadrupeds and birds upon the numerous obelisks brought from Egypt to ancient Rome; and all these figures, one with another, have a perfect resemblance to their intended objects, such as they still are in our days.

My learned colleague, M. Geoffroy Saint Hilaire, convinced of the importance of this research, carefully collected in the tombs and temples of Upper and Lower Egypt as many mummies of animals as he could procure. He has brought home the mummies of cats, ibises, birds of prey, dogs, monkies, crocodiles, and the head of a bull; and, after the most attentive and detailed examination, not the smallest difference is to be perceived between these animals and those of the same species which we now see, any more than between the human mummies and the skeletons of men of the present day. Some slight differences are discoverable between ibis and ibis, for example, just as we now find differences in the descriptions of naturalists; but I have removed all doubts on that subject, in a Memoir on the Ibis of the ancient Egyptians, in which I have clearly shewn that this bird is precisely the same in all respects at present that it was in the days of the Pharaohs.* I am aware that in these I only cite the monuments of two or three thousand years back; but this is the most remote antiquity to which we can resort in such a case.

* In that dissertation, the ibis of the ancient Egyptians is shewn to be a species of *numenius* or curlew, denominated by Cuvier *numenius ibis;* the same bird described in Bruce's Travels under the name of *abu-hannes.—Transl.*

From all these well established facts, there does not seem to be the smallest foundation for supposing, that the new genera which I have discovered or established among extraneous fossils, such as the *palæotherium*, *anoplotherium*, *megalonyx*, *mastodon*, *pterodactylis*, &c. have ever been the sources of any of our present animals, which only differ so far as they are influenced by time or climate. Even if it should prove true, which I am far from believing to be the case, that the fossil elephants, rhinoceroses, elks, and bears, do not differ farther from the presently existing species of the same genera, than the present races of dogs differ among themselves, this would by no means be a sufficient reason to conclude that they were of the same species; since the races or varieties of dogs have been influenced by the trammels of domesticity, which these other animals never did, and indeed never could experience.

Farther, when I endeavour to prove that the rocky strata contain the bony remains of several genera, and the loose strata those of several species, all of which are not now existing animals on the face of our globe, I do not pretend that a new creation was required for calling our present races of animals into existence. I only urge that they did not anciently occupy the same places, and that they must have come from some other part of the globe. Let us suppose, for instance, that a prodigious inroad of the sea were now to cover the continent of New Holland with a coat of sand and other earthy materials; this would necessarily bury the carcases of many animals belonging to the genera of *kanguroo*, *phascoloma*, *dasyurus*, *peramela*, *flying-phalangers*, *echidna*, and *ornithorynchus*, and would consequently entirely extinguish all the species of all these genera, as not one of them is to be found in any other country. Were the same revolution to lay dry the numerous narrow straits which separate New Holland from New Guinea, the Indian islands, and the continent of Asia, a road would be opened for the elephants, rhinoceroses, buffaloes, horses, camels, tigers, and all the other Asiatic animals, to occupy a land in which they are hitherto unknown. Were some future naturalist, after becoming well acquainted with the living animals of that country in this supposed new condition, to search below the surface on which these animals were nourished, he would then discover the remains of quite different races.

What New Holland would then be, under these hypothetical circumstances, Europe, Siberia, and a large portion of America, actually now are. Perhaps hereafter, when other countries shall be investigated, and New Holland among the rest, they also may be found to have all undergone similar revolutions, and perhaps may have made reciprocal changes of animal productions. If we push the former supposition somewhat farther, and, after the supply of Asiatic animals to New Holland, admit that a subsequent catastrophe might overwhelm Asia, the primitive country of the migrated animals, future geologists and naturalists would perhaps be equally at a loss to discover whence the then living animals of New Holland had come, as we now are to find out the original habitations of our present fossil animals.

§ 32. *Proofs that there are no Human Bones in a Fossil State.*

I now proceed to apply the previous reasonings to the human race. It is quite undeniable that no human remains have been hitherto discovered among the extraneous fossils; and this furnishes a strong proof that the extinct races which are now found in a fossil state, were not varieties of known species, since they never could have been subjected to human influence.

When I assert that human bones have not been hitherto found among extraneous fossils, I must be understood to speak of fossils or petrifactions, properly so called: As in peat depositions or turf bogs, and in alluvial formations, as well as in ancient burying-grounds, the bones of men with those of horses, and other ordinary existing species of animals, may readily enough be found; but among the fossil *palæotheria*, the elephants, the rhinoceroses, &c. the smallest fragment of human bone has never been detected. Most of the labourers in the gypsum quarries about Paris are firmly persuaded that the bones they contain are in a great part human; but after having seen and carefully examined many thousands of these bones, I may safely affirm that not a single fragment of them has ever belonged to our species.

I carefully examined at Pavia the collection of extraneous fossil bones brought there by Spallanzani from the island of Cerigo; and, notwithstanding the assertion of that celebrated observer, I affirm that there is not a single fragment among them that ever formed part of a human skeleton.

In my fourth volume, the *homo diluvii testis* of Scheuchzer is restored to the *proteus*, its true genus; and in a still more recent examination of it at Haerlem, allowed me by the politeness of M. Van Marum, who even permitted me to uncover some parts that were before enveloped in the stone, I obtained decisive proof of what I had before announced.

Among the fossil bones discovered at Cronstadt, the fragment of a jaw, together with some articles of human manufacture, was found; but it is well known that the ground was dug up without any precautions, and no notes were taken of the different depths at which each article was found. Everywhere else, the fragments of bone considered as human have been found to belong to some animal, either when the fragments themselves have been actually examined, or even when their engraved figures have been inspected. Such real human bones as have been found in a fossil state, belonged to bodies which had fallen into crevices of rocks, or had been left in the forsaken galleries of ancient mines, and were covered up by incrustation. The same has been the case with all articles of human fabric. The pieces of iron which have been found at Montmartre, are fragments of the iron tools used in the quarries for putting in blasts of gunpowder, and which sometimes break in the stone.

Yet human bones preserve equally well with those of animals, when placed in the same circumstances; and there is no observable difference in this respect in Egypt, between the mummies of men and those of quadrupeds. I have picked up, from the excavations made lately in the ancient church of St Genevieve, human bones that had been interred below the remains of the first race, which may even have belonged to some princes of the family of Clovis, and which still retained their forms very perfectly.* We do not find in ancient fields of battle, that the skeletons of men are more wasted than those of horses, except in so far as they may be influenced by size;

* M. Fourcroy has given an analysis of these bones.

and we find among extraneous fossils the bones of animals as small as rats, perfectly well preserved.

Every circumstance, therefore, contributes to establish this position—That the human race did not exist in the countries in which the fossil bones of animals have been discovered, at the epoch when these bones were covered up; as there cannot be a single reason assigned why men should have entirely escaped from such general catastrophes; or, if they also had been destroyed and covered over at the same time, why their remains should not be now found along with those of the other animals. I do not presume, however, to conclude that man did not exist at all before these epochs. He may have then inhabited some narrow regions, whence he went forth to repeople the earth after the cessation of these terrible revolutions and overwhelmings. Perhaps even the places which he then inhabited may have been sunk into the abyss, and the bones of that destroyed human race may yet remain buried under the bottom of some actual seas; all except a small number of individuals who were destined to continue the species.

However this may have been, the establishment of mankind in those countries in which the fossil bones of land animals have been found, that is to say, in the greatest part of Europe, Asia, and America, must necessarily have been posterior not only to the revolutions which covered up these bones, but also to those other revolutions, by which the strata containing the bones have been laid bare. Hence it clearly appears, that no argument for the antiquity of the human race in those countries can be founded either upon these fossil bones, or upon the more or less considerable collections of rocks or earthy materials by which they are covered.

§ 31. *Proofs of the recent Population of the World, and that its present Surface is not of very ancient Formation.*

On the contrary, by a careful investigation of what has taken place on the surface of the globe, since it has been laid dry for the last time, and its continents have assumed their present form, at least in such parts as are somewhat elevated above the level of the ocean, it may be clearly seen that this last revolution, and consequently the establishment of our existing societies, could not have been very ancient. This result is one of the best established, and least attended to, in rational zoology; and it is so much the more valuable, as it connects natural and civil history together in one uninterrupted series.

When we endeavour to estimate the quantity of effects produced in a given time by any causes still acting, by comparing them with the effects which these causes have produced since they began to operate, we may determine nearly the period at which their action commenced; which must necessarily be the same period with that in which our continents assumed their presently existing forms, or with that of the last retreat of the waters. It must have been since that last retreat of the waters, that the acclivities of our mountains have begun to disintegrate, and to form slopes or taluses of the debris at their bottoms and upon their sides; that our rivers have begun to flow in their present courses, and to form alluvial depositions; that our existing vegetation has begun to extend itself, and to form vegetable soil; that our present cliffs, or steep sloping coasts, have begun to be worn away by the waters of the sea; that our actual downs, or sand-hills, have begun to be blown up by the winds. And, dating from the same epoch, colonies of the human race must have then begun, for the first or for the second time, to spread themselves, and to form new establishments in places fitted by nature for their reception.

I do not here take the action of volcanoes into the account, not only because of the irregularity of their eruptions, but because we have no proofs of their not having been able to act below the sea; and because, on that account they cannot serve us as a measure of the time which has elapsed since its last retreat.

MM. Deluc and Dolomieu have most carefully examined the progress of the formation of new grounds by the collection of slime and sand washed down by the rivers; and, although exceedingly opposed to each other on many points of the theory of the earth, they agree exactly on this. These formations augment very rapidly; they must have increased with the greatest rapidity at first, when the mountains furnished the greatest quantity of materials to the rivers, * and

* One instance will be found appended to this Essay, of modern alluvial formations proceeding with considerably increas-

yet their extent still continues to be extremely limited.

The memoir by M. Dolomieu respecting Egypt,* tends to prove that the tongue of land on which Alexander caused his famous commercial city to be built, did not exist in the days of Homer; because they were then able to navigate directly from the island of Pharos into the gulf afterwards called *Lacus Mareotis;* and that this gulf, as indicated by Menelaus, was between fifteen and twenty leagues in length. Supposing this to be accurate, it had only required the lapse of nine hundred years, from the days of Homer to the time of Strabo, to reduce matters to the situation described by this latter author, when that gulf was reduced to the state of a lake only six leagues long.

It is a more certain fact, that since that time a still greater change has taken place. The sands, which have been thrown up by the sea and the winds, have formed, between the isle of

---

ed rapidity, in the researches of M. Prony respecting the alluvial depositions at the mouths of the Po.—*Transl.*

* In the Journal de Physique, Vol. XLII.

Pharos and the scite of ancient Alexandria, an isthmus more than four hundred yards broad, on which the modern city is now built. These collections of sand have also blocked up the nearest mouth of the Nile, and have reduced the lake Mareotis almost to nothing; while in the course of the same period, the Nile has deposited alluvial formations all along the rest of the coast. In the time of Herodotus, the coast of the Delta extended in a straight line, and is even represented in that direction in the maps constructed for the geography of Ptolemy: But since then the coast has so far advanced as to have assumed a semicircular projection into the Mediterranean. The cities of Rosetta and Damietta, built on the sea coast less than a thousand years ago, are now two leagues distant from the sea.

We may learn in Holland and Italy, how rapidly the Rhine, the Po, and the Arno, since they have been confined within dikes, now elevate their beds, and push forward the alluvial grounds at their mouths towards the sea, forming long projecting promontories at their sides; and it may be concluded, from this assured fact, that these rivers have not required the lapse of many centuries to deposit the low alluvial plains through which they now flow.

Many cities, which were flourishing sea-ports in well-known periods of history, are now several leagues inland, and several have even been ruined by this change. The inhabitants of Venice at present find it exceedingly difficult to preserve the *lagunes*, by which that once celebrated city is separated from the continent of Italy, from filling up; and there can be no doubt that she will some day become united to the main land, in spite of every effort to preserve her insular situation.*

We learn from Strabo, that Ravenna stood among *lagunes* in the time of Augustus, as Venice does now; but Ravenna is now at the distance of a league from the sea. Spina had been originally built by the Greeks on the sea-coast; but in the time of Strabo the sea was removed to the distance of ninety stadia. This city has been long since destroyed. Adria, which gave name to the Adriatic, was, somewhat more than twenty

* See a Memoir on the Lagunes of Venice, by M. Forfait.

centuries ago, the chief port of that sea, from which it is now at the distance of six leagues. The Abbé Fortis has even produced strong evidence for believing that the Euganian hills may have been islands, at a period somewhat more remote.

M. de Prony, a learned member of the Institute, and inspector-general of bridges and highways, has communicated to me some very valuable observations, to explain the changes which have taken place on the flat shores usually denominated the *Littoral* of the Adriatic, and which will be found appended to this Essay. Having been directed by government to examine and report upon the precautions which might be employed for preventing the devastations occasioned by the floods of the Po, he ascertained that this river has so greatly raised the level of its bottom, since it was shut in by dikes, that its present surface is higher than the roofs of the houses in Ferrara. At the same time, the alluvial additions produced by this river have advanced so rapidly into the sea, that, by comparing old charts with the present state, the coast appears to have gained no less than fourteen thousand yards since the year 1604, giving an average of an hundred and eighty to two hundred feet* yearly; and in some places the average amounts to two hundred feet. The Adige and the Po are both at present higher than the intervening lands; and the only remedy for preventing the disasters which are now threatened by their annual overflowings, would be to open up new channels for the more ready discharge of their waters, through the low grounds which have been formed by their alluvial depositions.

Similar causes have produced similar effects along the branches of the Rhine and the Maese; owing to which all the richest districts of Holland have the frightful view of their great rivers held up by dikes, at the height of twenty or even thirty feet above the level of the land.

M. Wiebeking, director of bridges and highways in the kingdom of Bavaria, has given an excellent memoir upon this subject, so highly im-

---

* In the appended extract from the Memoir of M. Prony, the older average yearly increase is stated at 25 *metres*, or 82 English feet and a quarter of an inch; and the average of the last 200 years at 70 *metres*, or 229 feet 7 inches and 9-tenths yearly.—*Transl.*

portant to be known and understood thoroughly, both by the people and the government, in all countries liable to these changes. In this memoir he has demonstrated that all rivers are continually elevating the levels of their beds, more or less, according to circumstances.

This formation and increase of new grounds by alluvial depositions, proceeds with as much rapidity along the coasts of the North Sea as on those of the Adriatic. These additions can be easily traced in Friesland and Groningen, where the epoch of the first dikes, constructed by the Spanish governor, Gaspard Robles, is well known to have been in 1570. An hundred years afterwards, the alluvial depositions had added in some places three quarters of a league of new land on the outside of these dikes: And the city of Groningen, partly built upon the ancient soil which has no connection with the present sea, being a calcareous formation in which the same species of shells are found as in the coarse limestone formations near Paris, is only six leagues from the sea. Having been upon the spot, I can give my testimony to the facts already so well stated by M. Deluc in his Letters to the Queen of England. The same phenomenon is as distinctly observable all along the coasts of East Friesland, and the countries of Bremen and Holstein, as the period at which the new grounds were enclosed by dikes for the first time is perfectly well known, and the extent that has been gained since can be easily measured. These new alluvial lands, left by the sea and the rivers, are of astonishing fertility, and are so much the more valuable as the ancient soil of these countries, being mostly covered by barren heaths and peat-mosses, is almost incapable of cultivation; so that the alluvial lands alone produce subsistence for the many populous cities that have been built along these coasts since the middle age, and which probably might not have reached their present flourishing condition, without the aid of these rich grounds, which have been, as it were, created by the rivers, and to which they are continually making additions.

If the size which Herodotus attributed to the sea of Asoph, which he says was equal to the Euxine, * had been less vaguely indicated, and if we could certainly ascertain what he understood

* Melpomene, LXXXVI.

to be the *Gerrhus*,* we should there find strong additional proofs of the great changes produced by the rivers, and of the rapidity with which these have been made. The alluvial depositions of these rivers, in the course of 2250 years since the time of Herodotus, have reduced the sea of Asoph to its present comparatively small size; have shut up entirely that branch of the Dneiper which formerly joined the *Hypacyris*, and discharged its waters along with that river into the gulf called *Carcinites*, now the *Olu-Degnitz*; and have now almost reduced the *Hypacyris* and the *Gerrhus* to nothing.†

We should possess proofs no less strong of the same thing, could we be certain that the Oxus or Sihon, which flows at present into Lake Aral, formerly reached the Caspian sea: But the

* Melpomene, LVI.

† See the Geography of Herodotus by M. Rennel, and the Physical Geography of the Black Sea, &c. by M. Dureau de la Malle.

In the latter work, p. 170, M. Dureau supposes Herodotus to have said that the Boristhenes and the Hypanis flowed into the Palus Meotis: But Herodotus, in Melpomene, LIII. only says that these two rivers discharged their waters into the same marsh; that is, into the Liman, exactly as in the present day; and Herodotus does not carry the Gerrhus and the Hypacyris any farther.

proofs which we possess on all these points are too vague, and even contradictory, to be admitted in support of physical propositions; and besides we are in possession of facts sufficiently conclusive, without being under the necessity of having recourse to those which are doubtful.

The downs or sand-hills which are thrown up by the sea upon low flat coasts, when the bed of the sea happens to be composed of sand, have been already mentioned. Wherever human industry has not succeeded to fix these downs, they advance as surely and irresistibly upon the land, as the alluvial formations from the rivers encroach upon the sea. In their progress inland, they push before them great pools of water, formed by the rain which falls on the neighbouring grounds, and which has no means of running off in consequence of the obstructions interposed by the downs. In several places these proceed with a frightful rapidity, overwhelming forests, houses, and cultivated fields, in their irresistible progress. Those upon the coast of the Bay of Biscay * have overwhelmed a great number of

* See Report respecting the Downs of the Gulf of Gascony, or Bay of Biscay, by M. Tassin, *Mont-de-Marsan*, *an.* X.

villages, which are mentioned in the records of the middle age; and even at present, in the single department of *Landes*, they threaten no fewer than ten with almost inevitable destruction. One of these, named Mimigan, has been in danger for the last fifteen years from a sand-hill of more than sixty feet in perpendicular height, which obviously continues to advance.

In the year 1802, the pools overwhelmed five fine farm-houses belonging to the village of St. Julian.* They have long covered up an ancient Roman road, leading from Bourdeaux to Bayonne, and which could still be seen about thirty years ago, when the waters were lower than they are now.† The river Adour, which is formerly known to have passed Old Boucat to join the sea at Cape Breton, is now turned to the distance of more than two thousand four hundred yards.

The late M. Bremontier, inspector of bridges and highways, who made several extensive works to endeavour to stop the progress of these downs,

---

* Memoir on the Means of fixing the Downs, by M. Bremontier.

† Report of M. Tassin, formerly cited.

estimated their progress at sixty feet yearly, and in some places at seventy-two feet. According to this calculation, it would require two thousand years to enable them to arrive at Bourdeaux; and, on the same data, they have taken somewhat more than four thousand years to reach their present situations. *

The *turbaries*, or peat-mosses, which have been formed so generally in the northern parts of Europe, by the accumulation of the remains of *sphagnum* and other aquatic mosses, afford another means of estimating the time which has elapsed since the last retreat of the sea from our present continents. These mosses increase in height in proportions which are determinate in regard to each. They surround and cover up the small knolls upon which they are formed; and several of these knolls have been covered over within the memory of man. In other places, the mosses gradually descend along the vallies, extending downwards like the *glaciers*; but these latter melt every year at their lower edges, while the mosses are not stopped by any thing what-

* Memoir of M. Bremontier.

ever in their regular increase. By sounding their depth down to the solid ground, we may form some estimate of their antiquity; and it may be asserted respecting these mosses, as well as respecting the downs, that they do not derive their origin from an indefinitely ancient epoch.

The same observations may be made in regard to the slips, or fallings, which sometimes take place at the bottom of all steep slopes in mountainous regions, and which are still very far from having covered these over. But as no precise measures of their progress have hitherto been applied, we shall not insist upon them at any greater length.

§ 32. *Proofs, from Traditions, of a great Catastrophe, and subsequent Renewal of Human Society.*

From all that has been said, it may be seen that nature every where distinctly informs us that the commencement of the present order of things cannot be dated at a very remote period; and it is very remarkable, that mankind every where speak the same language with nature, whether we consult their natural traditions on this subject, or consider their moral and political state, and the intellectual attainments which they had made at the time when they began to have authentic historical monuments. For this purpose we may consult the histories of nations in their most ancient books, endeavouring to discover the real facts which they contain, when disengaged from the interested fictions which often render the truth obscure.

The Pentateuch has existed in its present form at least ever since the separation of the ten tribes under Jeroboam, since it was received as authentic by the Samaritans as well as by the Jews; and this assures us of the actual antiquity of that book being not less than two thousand eight hundred years.* Besides this, we have no reason to doubt of the book of Genesis having been composed by Moses, which adds five hundred years to its antiquity.

Moses and his people came out of Egypt, which is universally allowed by all the nations

* Introduction to the Books of the Old Testament, by Eichhorn.—Leipsic, 1803.

of the west to have been the most anciently civilized kingdom on the borders of the Mediterranean. The legislator of the Jews could have no motive for shortening the duration of the nations, and would even have disgraced himself in the estimation of his own, if he had promulgated a history of the human race contradictory to that which they must have learnt by tradition in Egypt. We may therefore conclude, that the Egyptians had at this time no other notions respecting the antiquity of the human race than are contained in the book of Genesis. And, as Moses establishes the event of an universal catastrophe, occasioned by an irruption of the waters, and followed by an almost entire renewal of the human race, and as he has only referred it to an epoch fifteen or sixteen hundred years previous to his own time, even according to those copies which allow the longest interval, it must necessarily have occurred rather less than five thousand years before the present day.*

The same notions seem to have prevailed in Chaldea on this subject; as Berosus, who wrote

* Joseph. Antiq. Jud. lib. I. cap. 3.—Eusebii, Praep. Evang. lib. IX. cap. 4.—Syncelli, Chronogr.

at Babylon in the time of Alexander, speaks of the deluge nearly in the same terms with Moses, and supposes it to have happened immediately before Belus, the father of Ninus. *

Whatever may be the authenticity of the writings attributed to Sanconiatho, he does not appear to have mentioned the Deluge in his History of Phœnicia. † Yet this event seems to have been believed in Syria, as they shewed in the temple of Hierapolis, at a period indeed long after, the abyss through which they pretended that its waters had run off. ‡

Even in Egypt this tradition appears to have been forgotten, as we do not find any traces of it in the most ancient remaining fragments from that country. All of these indeed are posterior to the devastations committed by Cambyses;

---

* Eusebii, Præp. Ev. lib. I. cap. 10.

† The Deluge, according to the Hebrew text of the Scriptures, took place 2348 years before the commencement of the Christian era, or 4160 years before the present year 1813. The creation of the world, on the same authority, was 5817 years ago; but the Samaritan text extends that event to the distance of 6513 years, and the Septuagint to 7685 years.—*Transl.*

‡ Lucian, de Dea Syria.

and the little agreement there is among them sufficiently proves that they had been derived from mutilated fragments: For we cannot establish the smallest probable conformity between the lists of the kings of Egypt, as given by Herodotus in the era of Artaxerxes, by Erastosthenes and Manetho under the Ptolemies, and by Diodorus in the reign of Augustus; neither do they agree among themselves in the extracts which they pretend to have taken from the writings of Manetho.* Yet the Egyptian mythology seems to allude to these great events in the fabulous adventures of Typhon and Osiris. Besides, if the priests of Sais really gave the accounts to Solon, which are repeated by Critias in the writings of Plato, we must conclude that they had preserved some very exact traditions of a great revolution, though they had removed its epoch much farther back than was done by Moses. They had even theoretically devised a series of alternate revolutions; one set occasioned by means of water, and the other by means of fire; which notion had also prevailed among the Assyrians, and even in Etruria.

---

* See the English Ancient Universal History, Vol. I.

The Greeks, who derived their civilization at a late period from Phœnicia and Egypt, mixed the confused ideas which they had received of the mythologies of these nations with the equally confused vestiges of their own earliest history. The sun, personified under the name of Ammon, or the Egyptian Jupiter, was converted into a prince of Crete. *Phta*, the grand artisan or creator of all things, was converted into Hephestes, or Vulcan, a smith of Lemnos. *Chom*, another symbol of the sun, or of the divine power, was transformed into Heracles, or Hercules, a prodigiously strong hero of Thebes. The cruel *Moloch* of the Phœnicians, the same with the *Remphah* of the Egyptians, became with them *Chronos*, or Time, who devoured his own children, and was afterwards metamorphosed into Saturn, King of Italy.* When any violent inundation took place during the reign of any of their princes, the Greeks afterwards described it with

---

* See Jablonsky, Pantheon Ægyptiacum, and Gatterer, de Theogonia Egyptiorum, in the seventh volume of the Gottingen Memoirs.

These two authors do not agree, any more than the ancients, as to the significations of the Egyptian divinities; but they perfectly agree with each other, and with the ancient writers, as to the gross alterations made respecting them by the Greeks.

all the circumstances which had been handed down to them by tradition respecting the great deluge; and they represented Deucalion as having repeopled the earth, yet allowed a lengthened posterity to his uncle Atlas.

The incoherence of all these traditionary tales, while they attest the barbarism and ignorance of all the tribes around the Mediterranean, attest also the recentness of their establishments; and this very circumstance is in itself a strong proof of the existence of a great catastrophe. The Egyptians, it is true, spoke of hundreds of centuries, but these were filled by a succession of gods and demi-gods; and it is in a great degree ascertained in modern times, that the long series of years and of successive human kings which they placed after the demi-gods, and before the usurpation of the shepherds, belonged only to the successions of contemporaneous chiefs of several small states, instead of a single series of successive kings of all Egypt.

Macrobius * assures us that collections of observations of eclipses made in Egypt were pre-

* In Somnio Scipionis, 21.

served, which presupposed uninterrupted labour for at least twelve hundred years before the reign of Alexander. How comes it then, had this been the case, that Ptolemy should not have availed himself of any of these observations, though made in the country where he wrote?

There was no great empire as yet established in Asia at the time of Moses. Even the Greeks, notwithstanding their ingenuity in inventing fables, did not pretend even to invent an antiquity for their own nation; for the most ancient colonies from Egypt and Phœnicia, by which they were reclaimed from a state of barbarism, are not carried back more than four thousand years from the present era; and the most ancient authors in which these colonies are mentioned, are a thousand years posterior to the events. The Phœnicians themselves had only been recently established in Syria, when they began to form establishments in Greece.

The astronomical observations of the Chaldeans, sent by Calisthenes to Aristotle, are said to have gone back for a period of four thousand years, if Simplicius is to be credited, who reports the story six hundred years after Aristotle. But the authenticity of this is exceedingly doubtful, as the Chaldean observations of eclipses actually preserved and cited by Ptolemy, do not go back more than two thousand five hundred years. * At all events, the Babylonian, or first Assyrian empire, could not have been long powerful, as there remained all around many unsubjected tribes, such as all those of Syria, until after the establishment of what is called the Second Kingdom of Assyria. The thousands of years therefore which the Chaldeans assumed, must have been equally fabulous with those of the Egyptians; or rather may be considered as astronomical periods, calculated backwards upon the basis of inaccurate observations; or merely as imaginary and arbitrary cycles, multiplied into themselves.†

The most reasonable among the ancients were of the same opinion, and have only carried back the reigns of Ninus and Semiramis, the earliest of the conquerors, a little more than four thou-

---

* It is not quite obvious, from the language of the author, whether these are meant as pointing backwards from the respective epochs of Aristotle and Ptolemy, or only from the present day: the latter must, however, be the case.—*Transl.*

† See Memoire of M. de Guignes in the Acad. des Belles Lettres, Tom. XLVII. and the Voyage of M. Gentil. I. 241.

sand years. After them history continues long silent, * whence it may even be strongly suspected that these were only late inventions of the historians.

Our existing civilization and learning have been uninterruptedly transmitted down to us from the Egyptians and Phœnicians, through the Greeks and Romans; and we have derived immediately from the Jews our more pure ideas of morals and religion. Some small portions of knowledge have also come down to us from the Jews and Greeks, which they had derived variously from the Chaldeans, the Persians, and the Indians; and it is a most remarkable circumstance, that all these nations form only one original race, resembling each other in their physiognomies, and even in many conventional matters, such as their divinities, the names of the constellations, and even in the roots of their languages. †

* See Velleius Paterculus and Justin.

† For the analogy of the languages of India, Persia, and our western world, see the Mithridates of Adelung. On the analogy of the deities of the Indians, Egyptians, Greeks, and Romans, consult the works of Jablonsky and Gatterer, already

The Hindoos, perhaps the most anciently civilized people on the face of the earth, and who have least deviated from their originally established forms, have unfortunately no history. Among an infinite number of books of mystical theology and abstruse metaphysics, they do not possess a single volume that is capable of affording any distinct account of their origin, or of the various events that have occurred to their communities. Their *Maha-Bharata*, or pretended great history, is nothing more than a poem. The *Pouranas* are mere legends; on comparing which with the Greek and Latin authors, it is excessively difficult to establish a few slight coincidences of chronology, and even that is continually broken off and interrupted, and never goes back farther than the time of Alexander. *

---

cited; as also the Memoir of Sir William Jones, with the notes of M. Langlès, in the first volume of the French translation of the Calcutta Memoirs, p. 192, *et seq.* The identity of the constellations, especially of the signs of the Zodiac, between the Hindoos and the most western nations, with the names given to the days of the week, &c. are now universally known.

* Consult the elaborate Memoir of M. Paterson, respecting the kings of Magadaha, emperors of Hindostan, and upon the epochs of Vicramadityia and Salahanna, in the Calcutta Memoirs, Vol. IX.

It is now clearly proved that their famous astronomical tables, from which it has been attempted to assign a prodigious antiquity to the Hindoos, have been calculated backwards ; * and it has been lately ascertained, that their *Surya-Siddhanta*, which they consider as their most ancient astronomical treatise, and pretend to have been revealed to their nation more than two millions of years ago, must have been composed within the seven hundred and fifty years last past. † Their *Vedas*, or sacred books, judging from the calendars which are conjoined with them, and by which they are guided in their religious observances, and estimating the *colures* indicated in these calendars, may perhaps go back about three thousand two hundred years, which nearly coincides with the epoch of Moses. ‡ Yet the Hindoos are not entirely ignorant of the revolutions which have affected the globe, as their theology has in some measure consecrated

---

* See Expos. du Syst. du Monde, by M. de la Place p. 330.

† See the Memoir by M. Bentley, on the Antiquity of the Surya-Siddhanta, in the Calcutta Memoirs, Vol. VI. p. 537, and the Memoir by the same Author on the Astronomical Systems of the Hindoos, *ibid.* Vol. IX. p. 195.

‡ See the Memoir by M. Colebrooke upon the Vedas, and particularly p. 493, in the Calcutta Memoirs, Vol. VIII.

certain successive destructions which its surface has already undergone, and is still doomed to experience; and they only carry back the last of those, which have already happened, about five thousand years; * besides which, one of these revolutions is described in terms nearly corresponding with the account given by Moses. † It is also very remarkable, that the epoch at which they fix the commencement of the reigns of their

---

* Voyage to India by M. Le Gentil. I. 235.—Bentley in the Calcutta Memoirs, Vol. IX. p. 222.—Paterson, in ditto. *ibid.* p. 86.

† Sir William Jones, in the Calcutta Memoirs, French translation, Vol. I. p. 170.

The English reader may be gratified by the following extract from this dissertation of Sir William Jones.—*Transl.*

"We may fix the time of Buddah, or the ninth great incarnation of Vishnu, in the year 1014 before the birth of Christ. The Cashmirians, who boast of his descent in their kingdom, assert that he appeared on earth about two centuries after Chrishna, the Indian Apollo....................We have therefore determined another interesting epoch, by fixing the age of Crishna near the year 1214 before Christ. As the three first *avatars* or descents of Vishnu, relate no less clearly to an *Universal Deluge*, in which eight persons only were saved, than the fourth and fifth do to the punishment of impiety and the humiliation of the proud; we may for the present assume that the second, or silver age of the Hindus, was subsequent to the dispersion from Babel; so that we have only a dark interval of about a thousand years, which were employed in the settlement of nations, and the cultivation of civilized society."—*Works of Sir William Jones*, I. 29, 4to. *London*, 1799.

first human sovereigns, of the race of the Sun and Moon, is nearly the same at which the ancient authors of the west have placed the origin of the Assyrian monarchy, or about four thousand years ago.

It were quite in vain to attempt looking for any indications of these great events among the people of more southern regions, such as the Arabians or Abyssinians, as their ancient books are no longer existing; and the only histories they possess relative to remote antiquity are of recent compilation, and have been modelled after our Bible: hence all that their books contain respecting the deluge is borrowed from Genesis, and does not contribute any support to its authority. The Guebres, however, or Parsis, who are now the sole depositaries of the doctrines of Zoroaster and the ancient Persians, speak also of an universal deluge as having happened before the reign of *Cayoumarats*, their first king.

In order to recover some truly historical traces of the last grand *cataclysma*, or universal deluge, we must go beyond the vast deserts of Tartary, where, in the north-east of our ancient continent, we meet with a race of men differing entirely from us, as much in their manners and customs, as they do in their form and constitution. Their oral language is entirely monosyllabic, and they use arbitrary hieroglyphics instead of writing. They only possess a system of political morals, without any established religion; as the superstitions of the sect of Fo have been imported by them from India. Their yellow skins, high cheekbones, narrow and oblique eyes, and thinly scattered beards, give them an appearance so entirely different from us, that one is almost tempted to suspect that their ancestors and ours had escaped from the last grand catastrophe at two different sides: but, however this may have been, they date their deluge nearly at the same period with ours.

The *Chou-King*,* the most ancient of the Chinese books, is said to have been compiled by Confucius, about two thousand five hundred years ago, from fragments of more ancient works. Two hundred years afterwards, under the Emperor Chi-hoang-ti, the men of letters were persecuted, and all books were destroyed. About

* See the preface to the translation of the Chou-King, by M. de Guignes.

forty years after this persecution, an old *literati* restored a portion of the *Chou-King* from memory, and another portion was recovered that had been concealed in a tomb; but nearly the half was lost for ever. This, which is considered as the most authentic of all the Chinese books, begins the history of the country with an emperor named *Yao*, whom it represents as having let loose the waters, in the following terms: *Having raised himself to heaven, Yao bathed the feet even of the highest mountains, covered the less elevated hills, and rendered the plains impassable.* According to some accounts, the reign of Yao was four thousand five hundred years ago; while others only carry it back to three thousand nine hundred and thirty years before the present time.

The same book, only a few pages farther on, introduces one *Yu*, prime minister and chief engineer, re-establishing the courses of the rivers, building dykes, digging canals, and regulating the taxes of all the provinces of China, that is, of an empire which extends six hundred leagues in all directions. But the utter impossibility of such operations, immediately after such events, shews clearly that the whole story can only be considered as a moral and political romance.

More modern Chinese historians have introduced a long series of emperors before Yao, which they have combined with a multitude of fabulous circumstances, yet without venturing to assign any fixed dates to their reigns. These writers also continually differ from each other, both in the number and names of the kings; and none of them are universally approved on this subject by their countrymen.

The introduction of astronomy into China is attributed to Yao; but the real eclipses recorded by Confucius, in his Chronicle of the Kingdom of *Lou*, only go back two thousand six hundred years, hardly half a century higher than those of the Chaldeans, as related by Ptolemy. In the *Chou-King* indeed, there is an eclipse mentioned which goes back three thousand nine hundred and sixty-five years, but which is related with the addition of so many absurd circumstances, that it has been probably invented at a subsequent period. A conjunction also is stated as having happened four thousand two hundred and fifty-nine years ago, which would therefore be the most ancient known astronomical observation, but its authenticity is contested. The earliest observation that appears to rest upon good grounds, is one made by means of a gnomon, two thousand nine hundred years ago.

It is not to be conceived that mere chance should have thus given rise to so striking a coincidence between the traditions of the Assyrians, the Hindoos, and the Chinese, in attributing the origins of their respective monarchies so nearly to the same epoch, of about four thousand years before the present day. The ideas of these three nations, which have so few features of resemblance, or rather which are so entirely dissimilar in language, religion, and laws, could not have so exactly agreed on this point, unless it had been founded upon truth.

We do not require any specific dates from the natives of America, who were not possessed of any real writing, and whose most ancient traditions only go back a few centuries before the arrival of the Spaniards. Yet even among them some traces of a deluge are conceived to have been found in their barbarous hieroglyphics.*

* See the excellent and magnificent work of Humboldt, upon the monuments of the Mexicans.

The Negroes, the most degraded race among men, whose forms approach nearest to those of the inferior animals, and whose intellect has not yet arrived at the establishment of any regular form of government, nor at any thing which has the least appearance of systematic knowledge, have preserved no sort of annals or of tradition; and from them therefore we are not to expect any information on the subject of our present researches. Yet even the circumstances of their character clearly evince that they also have escaped from the last grand catastrophe, perhaps by another route than the races of the caucassan and altaic chains, from whom perhaps they may have been long separated before the epoch of that catastrophe.

Thus all the nations which possess any records or ancient traditions, uniformly declare that they have been recently renewed, after a grand revolution in nature. This concurrence of historical and traditionary testimonies, respecting a comparatively recent renewal of the human race, and their agreement with the proofs that are furnished by the operations of nature, which have been already considered, might certainly warrant us in refraining from the examination of certain equivocal monuments, which have been brought forward by some authors in support of a contrary opinion. But even this examination, to judge of it by some attempts already made, will probably do nothing else than add some more proofs to that which is furnished by tradition.

§ 33. *Proofs derived from several Miscellaneous Considerations.*

It does not now appear that the famous zodiac, in the porch of the temple at Dendera, can support the opinion which some have been disposed to deduce from it, respecting the high antiquity of the present race of mankind. Nothing can be drawn for this purpose from its division into two bands of six signs each, as indicative of the position of the colures produced by the precession of the equinoxes, or to show that these do not merely answer to the commencement of the civil year of the Egyptians at the period when it was drawn. As the civil year in Egypt consisted exactly of three hundred and sixty-five days, it made the tour of the zodiac in fifteen hundred and eight years; or, according to the Egyptians, which shows that they had not observed it in fourteen hundred and sixty years. In the same temple there is another zodiac, in which the sign *Virgo* is represented as beginning the year. If these circumstances were connected with the position of the solstice, this other zodiac in the interior of the temple must have been drawn two thousand years before that in the porch; but supposing it to represent the commencement of the civil year, an interval of very little more than an hundred years is quite sufficient to reconcile the two zodiacs with each other.

It may be inquired also, whether our zodiac may not contain some internal proofs of its antiquity, and whether the figures which have been employed to represent its signs or constellations, may not have some reference to the colures at the epoch when they were adopted. All, however, that has been advanced on this subject, is founded on allegories, supposed to be contained in the several figures. Thus it has been supposed, that *Libra*, or the balance, indicated the equality of the days and nights; *Taurus*, or the bull, the season of labouring the earth; *Cancer*, or the crab, a retrogradation of the sun; *Virgo*, the season of gathering in the fruits of the earth;

and so of the rest. All this is mere bold conjecture: But besides, these explanations must necessarily vary for every country; and it would be requisite to assign a different epoch to each separate zodiac, according to the climate of the country in which it is supposed to have been invented; nay, perhaps, there may be no climate and no epoch in conformity with which rational explanations could be devised for all the signs. It is also possible, that these names may have been given at a very remote period, without reference at all to the divisions of time or space, or to the different states of the sun in its course, just as they are now given by astronomers; and may have been applied to the constellations or groups of stars, as referring to a particular epoch merely by chance; so that nothing whatever can be deduced from their significations. *

It may be objected, that the advanced state of astronomy among these ancients is a striking

---

* See the dissertation by M. de Guignes respecting the zodiacs of the oriental nations, in the Memoirs of the Academy of Belles Letters, Vol. XLVII.

proof of their high antiquity, and that it must have required a vast many centuries of observations by the Chaldeans and Indians to enable them to acquire the knowledge which they certainly possessed nearly three thousand years ago, respecting the length of the year, the precession of the equinoxes, the relative motions of the sun and moon, and several other important circumstances. But to explain all this, without the necessity of any prodigious antiquity, it may be remarked, that a nation may well be expected to make rapid progress in any particular science that has no other to attend to; and that with the Chaldeans especially, the perpetual serenity and clearness of their sky, the pastoral life which they led,* and the peculiar superstition to which

---

* It may be here noticed, that our present shepherds have infinitely more practical knowledge of astronomy, merely from being so much in the open air, almost unemployed, than all the other ordinary ranks in society. An instance of astonishingly rapid progress in that science was exhibited in our own day by the celebrated James Ferguson, who constructed an accurate map of the heavens when a herd-boy, entirely from his own untutored genius. Had astronomy been then a non-existent science, even he might have carried it almost as far as the Chaldeans in a single lifetime; and perhaps, in mapping the heavens, he went farther even than all the astronomers of Chaldea.—*Transl.*

they were addicted, rendered the stars a general object of attention. They had also colleges, or societies of their most respectable men, appointed to make astronomical observations, and to put them upon record. Let us suppose, also, that among so many persons who had nothing else to do, there were two or three possessed of singular talents for the study of geometrical science, and every thing known to that people might easily have been accomplished in a very few centuries.

Since the time of the Chaldeans, real astronomy has only had two eras; that of the Alexandrian school, which lasted four centuries, and that of our own times, which has not yet lasted so long. The learned period of the Arabs hardly added any thing to that science, and all the other ages of the world were mere blanks with respect to it. Three hundred years did not intervene between Copernicus and De la Place, the celebrated author of the *Mécanique Céleste*; yet some wish to believe that the Hindoos must have had many thousand years to discover their astronomical rules. After all, even were every thing that has been fancied respecting the antiquity of astronomy as fully proved as it appears to us destitute of proof, it would establish no conclusion against the great catastrophe, which has left in other respects so many convincing monuments of its own existence. All that it is necessary to admit, even on that supposition, is, what some moderns have thought,—That astronomy was among the number of the sciences that were preserved by the small number of men who escaped from that catastrophe.

The antiquity of certain mining operations has also been prodigiously exaggerated by some writers. A recent writer pretends that the mines of the island of Elba, to judge from their wastes, must have been explored above forty thousand years ago; while another author, who has also examined these wastes with much attention, reduces the interval to somewhat more than five thousand years, supposing that the ancients wrought out every year one-fourth only of the quantity that is wrought out in the present day. * We have no reason, however, to believe that the Romans, who consumed so much

* See History of China, before the Deluge of the Ogigians, by M. de Fortin d'Urban. II. 33.

iron in their armies, were so slow in their mining operations as this high antiquity of the mines of Elba would imply; and besides, even if these mines had been wrought for no more than four thousand years, how should it have been that iron was so little known among the ancients in the first ages of Greece and Rome?

§ 34. *Concluding Reflections.*

I am of opinion, then, with M. Deluc and M. Dolomieu,—That, if there is any circumstance thoroughly established in geology, it is, that the crust of our globe has been subjected to a great and sudden revolution, the epoch of which cannot be dated much farther back than five or six thousand years ago; that this revolution had buried all the countries which were before inhabited by men and by the other animals that are now best known; that the same revolution had laid dry the bed of the last ocean, which now forms all the countries at present inhabited; that the small number of individuals of men and other animals that escaped from the effects of that great revolution, have since propagated and spread over the lands then newly laid dry; and consequently, that the human race has only resumed a progressive state of improvement since that epoch, by forming established societies, raising monuments, collecting natural facts, and constructing systems of science and of learning.

Yet farther,—That the countries which are now inhabited, and which were laid dry by this last revolution, had been formerly inhabited at a more remote era, if not by man, at least by land animals; that, consequently, at least one previous revolution had submerged them under the waters; and that, judging from the different orders of animals of which we discover the remains in a fossil state, they had probably experienced two or three irruptions of the sea.

These alternate revolutions form, in my opinion, the problem in geology that is most important to be solved, or rather to be accurately defined and circumscribed; for, in order to solve it satisfactorily and entirely, it were requisite that we should discover the cause of these events, —an enterprise involving difficulties of a very different nature.

We are able to discover with sufficient precision all that takes place on the surface of our world in its present state, and we have sufficiently ascertained the uniform progress and regular successions of the primitive formations; but the study of the secondary formations is as yet scarcely commenced. The wonderful series of unknown marine moluscæ and zoophites, followed by fossil remains of serpents and of fresh-water fish equally unknown, which are again succeeded by other moluscæ and zoophites more nearly allied to those which exist at present: All these land animals, these moluscæ, and other unknown animals of fresh water, which next occupy the formations, and which are finally succeeded by other moluscæ and other animals resembling those of our present seas; the relations between these various animals and the plants whose remains are mixed among them, and the relations of both with the mineral strata in which they are imbedded; the little resemblance between these extraneous fossils of animals and plants, as contained in the different basins of former waters:—All these form a series of phenomena which imperiously demands the attention of philosophers.

This study is rendered interesting, by the variety of productions of partial or general revolutions which it affords, and by the abundance of the different species which alternately offer themselves to view; it neither has that dull monotony which attaches to the study of the primitive formations, nor does it force us, like the latter, almost necessarily into hypotheses. The facts with which it is conversant are so prominent, so curious, and so obvious, that they may suffice to occupy the most ardent imagination; and the conclusions which they afford from time to time, even to the most cautious observer, have nothing vague or arbitrary in their nature. Finally, by the careful investigation of these events, which approach, as it were, to the history of our own race, we may hope to be able to discover some traces of more ancient events and their causes; if, after so many abortive attempts already made on the same subject, we may yet flatter ourselves with that hope.

These ideas have haunted, and I may even say, have tormented me, during all my researches into the fossil remains of bones, of which I now offer the results to the public; and though these only contain a very small portion of the phenomena connected with the immediately preceding period of the history of the earth, they yet connect themselves most intimately with all the rest. It was hardly possible to avoid endeavouring to examine these phenomena in the country immediately round Paris; and my excellent friend M. Brongniart, led by other studies to have similar views, associated himself with me in the investigation, by which we laid the foundation of our Essay on the Mineral Geography of Paris. That work, however, although it bears my name, has become almost entirely the work of my friend, in consequence of the infinite care he has bestowed, ever since the first conception of our plan, and during the execution of our several surveys and researches, in the thorough investigation of all the objects of our research, and in the composition of the Essay itself.

The Essay on the Mineral Geography of the Environs of Paris, affords the most complete and satisfactory evidences of the principal facts and circumstances which I have endeavoured to establish in this discourse. It contains a history of the most recent changes which have taken place in one particular basin, and leads us as far as the chalk formation, which is infinitely more extended over the globe than the formations composed of those materials which are found in the basin of Paris. The chalk formation, which was before conceived to be of very modern origin, has been shewn in that extensive examination to have originated at a period considerably far back in the age before the last; or, in other words, to have owed its origin to causes connected with the revolution and catastrophe before the last general irruption of the waters over our present habitable world.

It would now be of great importance to examine the other basins containing chalk formations, and in general to pay particular attention to the strata which rest upon that formation, that these may be compared with those we found in the environs of Paris. Perhaps the chalk itself may be found to contain some successive depositions of organic remains. It is surrounded and supported by the compact limestone, which occupies a great proportion of France and Germany, and the extraneous fossils of which are extremely different from all those of our basin. But, in following the compact limestone, from the chalk to the limestone of the central ridges of Jura, which are almost devoid of shells, or to the aggregated rocks of the acclivities of the Hartz, the Vosges, and the Black Forest, we shall probably find abundance of variations: And the gryphites, the cornua ammonis, and the interochi, with which it abounds, may perhaps be found distributed by genera, or at least by species.

This compact limestone formation is not every where covered over by chalk. Without that intervening, it surrounds basins in several places, or supports elevated flats or table lands not less worthy of examination than those which are limited by chalk. We should derive great information, for instance, from a history of the gypsum quarries of Aix, in which, as well as in those of Paris, reptiles and fresh-water fishes are found; and probably land-animals will be also discovered by careful research; while we are assured that nothing similar occurs in the entire interval between these two places, which are almost two hundred leagues distant from each other.

The long ranges of sand-hills which skirt both slopes of the Appenines through almost the entire length of Italy, contain every where perfectly well-preserved shells, which are often found retaining their colours, and even their natural pearl-like polish, and several of which resemble those still found in our seas. It would be of great importance to be well acquainted with these, and to have all their successive strata accurately examined, determining the extraneous fossils found in each, and comparing them with those that are contained in other recent strata; such, for example, as those in the environs of Paris.

In the course of this investigation, it would be proper to connect the series, on the one hand, with the most solid and most ancient formations, and, on the other, with the recent alluvial depositions made by the Po, the Arno, and their tributary streams; as also, to determine their relations with the innumerable masses of volcanic productions which are interposed between them; and, finally, to ascertain the mutual situations of the various sorts of shells, and of the fossil bones of elephants, rhinoceroses, hippopotami, whales, cachalots, and dolphins, in which several of these hills abound. I have only a very superficial knowledge of these lower hills of the Appenine chain, acquired in the course of a journey devoted to other objects; but I am of opinion that they contain the true secret of the last operations of the sea.

There are many other strata, even celebrated for their extraneous fossils, which have not been hitherto so accurately examined as to enable them to be connected with the general series, and whose relative antiquity, therefore, has not been ascertained. The copper slate of Thuringia is said to be filled with the remains of fresh-water fish, and to be older than most of the secondary or flœtz formations. We are also as yet uninformed of the real position of the stinkstone slate of Oeningen, which is also said to be full of the remains of fresh-water fish; of that of Verona, evidently abounding in the remains of sea-fish, but which have been very improperly named by the naturalists who have described them; of the black slate of Glacis; of the white slate of Aichstedt, also filled with the remains of fishes, of crabs, and of other marine animals different from shells. All these desiderata have as yet received no satisfactory explanation in books of geology; neither has it been as yet explained, why shells should be found almost every where, while fish are confined only to a few places.

It appears to me, that a consecutive history of such singular deposits would be infinitely more valuable than so many contradictory conjectures respecting the first origin of the world and other planets, and respecting phenomena which have confessedly no resemblance whatever to those of the present physical state of the world; such conjectures finding, in these hypothetical facts, neither materials to build upon, nor any means of verification whatever. Several of our geologists resemble those historians who take no interest in the history of France, except as to what passed before the time of Julius Cæsar. Their imaginations, of course, must supply the place of authentic documents; and accordingly each composes his romance according to his own fancy. What would become of these historians, if they had not been assisted in their combinations by the knowledge of posterior facts? But our geologists neglect exactly those posterior geological facts, which might at least, in some measure, dispel the darkness of the preceding times.

It would certainly be exceedingly satisfactory to have the fossil organic productions arranged in chronological order, in the same manner as we now have the principal mineral substances. By this, the science of organization itself would be improved; the developements of animal life; the succession of its forms; the precise determinations of those which have been first called into existence; the simultaneous production of certain species, and their gradual extinction;—all these would perhaps instruct us fully as much in the essence of organization, as all the experiments that we shall ever be able to make upon living animals: And man, to whom only a short space of time is allotted upon the earth, would have the glory of restoring the history of thousands of ages which preceded the existence of the race, and of thousands of animals that never were cotemporaneous with his species.

END OF THE ESSAY.

SUPPLEMENT:

Being an Extract from the Researches of M. de Prony, on the Hydraulic System of Italy; Containing an Account of the Displacement of that Part of the Coast of the Adriatic which is occupied by the Mouths of the Po.

THAT portion of the shore of the Adriatic which lies between the lake, or rather *lagune*, of Commachio and the *lagunes* of Venice, has undergone considerable alterations since ancient times, as is attested by authors worthy of entire credit, and as is still evidenced by the actual state of the soil in the districts near the coast; but it is impossible now to give any exact detail of the successive progress of these changes, and more especially of their precise measures, during the ages which preceded the twelfth century of our era.

We are however certain, that the city of *Hatria,* now called *Adria,* was formerly situated on the edge of the coast; and by this we attain a known fixed point upon the primitive shore, whence the nearest part of the present coast, at the mouth of the Adige, is at the distance of 25,000 *metres*;* and it will be seen in the sequel, that the extreme point of the alluvial promontory, formed by the Po, is farther advanced into the sea than the mouth of the Adige by nearly 10,000 metres.†

The inhabitants of Adria have formed exaggerated pretensions, in many respects, as to the high antiquity of their city, though it is undeniably one of the most ancient in Italy, as it gave name to the sea which once washed its

---

* Equal to 27,340 yards and 10 inches English measure, or 15½ miles and 60 yards.

In these reductions of the revolutionary French *metres* to English measure, the *metre* is assumed as 39.37 English inches. —*Transl.*

† Or 10,936 yards and 4 inches, equal to 6 miles and nearly a quarter, English measure.

Hence the entire advance of the alluvial promontory of the Po appears to have extended to 21 miles 5 furlongs and 216 yards.—*Transl.*

walls. By some researches made in its interior and its environs, a stratum of earth has been found mixed with fragments of Etruscan pottery, and with nothing whatever of Roman manufacture. Etruscan and Roman pottery are found mixed together in a superior bed, on the top of which the vestiges of a theatre have been discovered. Both of these beds are far below the level of the present soil. I have seen at Adria very curious collections, in which these remains of antiquity are separately classed; and having some years ago observed to the viceroy, that it would be of great importance, both to history and geology, to make a thorough search into these buried remains at Adria, carefully noticing the levels in comparison with the sea, both of the primitive soil, and of the successive alluvial beds, his highness entered warmly into my ideas; but I know not whether these propositions have been since carried into effect.

Following the coast, after leaving Hatria, which was situated at the bottom of a small bay or gulf, we find to the south a branch of the *Athesis* or Adige, and of the *Fossa Philistina*, of which the remaining trace corresponds to what might have been the Mincio and Tartaro united, if the Po had still run to the south of Ferrara. We next find the *Delta Venetum*, which seems to have occupied the place where the lake or lagune of Commachio is now situated. This delta was traversed by seven branches of the *Eridanus* or Po, formerly called also the *Vadis Padus* or *Podincus;* which river, at the diramification of these seven branches, and upon its left or northern bank, had a city named *Trigoboli* whose scite could not be far from where Ferrara now stands. Seven lakes, inclosed within this delta, were called *Septem Maria*, and Hatria was sometimes denominated *Urbs Septem Marium*, or the city of the seven seas or lakes.

Following the coast from Hatria to the northwards, we come to the principal mouth of the *Athesis* or Adige, formerly named *Fossa Philistina*, and afterwards *Estuarium Altini*, an interior sea, separated, by a range of small islands, from the Adriatic gulf, in the middle of which was a cluster of other small isles, called *Rialtum*, and upon this archipelago the city of Venice is now seated. The *Estuarium Altini*, is what is now called the lagune of Venice, and no longer communicates with the sea, except by five passages, the small islands of the archipelago having been united into a continuous dike.

To the east of the lagunes, and north from the city of Este, we find the *Euganian* mounts or hills, forming, in the midst of a vast alluvial plain, a remarkable isolated group of rounded hillocks, near which spot the fable of the ancients supposes the fall of Phæton to have taken place. Some writers have supposed that this fable may have originated from the fall of some vast masses of inflamed matters near the mouths of the Eridanus, that had been thrown up by a volcanic explosion; and it is certain that abundance of volcanic products are found in the neighbourhood of Padua and Verona.

The most ancient notices that I have been able to procure respecting the situation of the shores of the Adriatic at the mouths of the Po, only begin to be precise in the twelfth century. At that epoch the whole waters of this river flowed to the south of Ferrara, in the *Po de Volano* and the *Po di Primaro*, branches which inclosed the space occupied by the *lagune* of Commachio. The two branches which were next formed by an irruption of the waters of the Po to the north of Ferrara, were named the river of *Corbolo*, *Langola*, or *Mazzorno*, and the river *Toi*. The former, and more northern of these, received the *Tartaro* or *canal bianco* near the sea, and the latter was joined at Ariano by another branch derived from the Po, called the *Goro* river. The sea-coast was evidently directed from south to north, at the distance of ten or eleven thousand *metres* * from the meridian of Adria; and *Loreo*, to the north of *Mesola*, was only about 2000 *metres* † from the coast.

Towards the middle of the twelfth century, the flood-waters of the Po were retained on their left or northern side by dikes near the small city of *Ficarolo*, which is about 19,000 *metres* ‡ to the north-west of Ferrara, spreading themselves southwards over the northern part of the territory of Ferrara and the *Polesine* of Rovigo, and flowed through the two formerly-mentioned canals of *Mazzorno* and *Toi*. It seems perfectly ascertained, that this change in the direction of

---

* Equal to 10,936 or 12,030 yards English measure.—*Transl.*

† Or 2,186 yards 2 feet English.—*Transl.*

‡ Or 20,778 yards 1 foot 10 inches.—*Transl.*

the waters of the Po had been produced by the effects of human labours; and the historians who have recorded this remarkable fact only differ from each other in some of the more minute details. The tendency of the river to flow in the new channels, which had been opened for the more ready discharge of its waters when in flood, continually increased; owing to which the two ancient chief branches, the *Voalno* and *Primaro*, rapidly decreased, and were reduced in less than a century to their present comparatively insignificant size; while the main direction of the river was established between the mouth of the Adige to the north, and what is now called *Porto di Goro* on the south. The two before-mentioned canals of *Mazzorno* and *Toi*, becoming insufficient for the discharge, others were dug; and the principal mouth, called *Bocco Tramontana*, or the northern mouth, having approached the mouth of the Adige, the Venetians became alarmed in 1604; when they excavated a new canal of discharge, named *Taglio de Porto Viro*, or *Po delle Fornaci*, by which means the *Bocco Maestra* was diverted from the Adige towards the south.

During four centuries, from the end of the

twelfth to that of the sixteenth, the alluvial formations of the Po gained considerably upon the sea. The northern mouth, which had usurped the situation of the *Mazzorno* canal, becoming the *Ramo di Tramontana,* had advanced in 1600 to the distance of 20,000 *metres* * from the meridian of Adria; and the southern mouth, which had taken possession of the canal of *Toi,* was then 17,000 *metres* † advanced beyond the same point. Thus the shore had become extended nine or ten thousand *metres* ‡ to the north, and six or seven thousand to the south. § Between these two mouths there was formerly a bay, or a part of the coast less advanced than the rest, called *Sacca di Goro.* During the same period of four hundred years previous to the commencement of the seventeenth century, the great and extensive embankments of the Po were constructed; and, also during the same period, the southern slopes of the Alps began to be cleared and cultivated.

---

* Or 21,872 yards.—*Transl.*

† Or 18,591 yards.—*Transl.*

‡ Equal to 9,842 or 10,936 yards.—*Transl.*

§ Equal to 6,564 or 7,655 yards.—*Transl.*

The great canal, denominated *Taglio di Porto Viro* or *Podelle Fornaci*, ascertains the advance of the alluvial depositions in the vast promontory now formed by the mouths or delta of the Po. In proportion as their entrances into the sea extend from the original land, the yearly quantity of alluvial depositions increases in an alarming degree, owing to the diminished slope of the streams, which was a necessary consequence of the prolongation of their bed, to the confinement of the waters between dikes, and to the facility with which the increased cultivation of the ground enabled the mountain torrents which flowed into them to carry away the soil. Owing to these causes, the bay called *Sacca di Goro* was very soon filled up, and the two promontories which had been formed by the two former principal mouths of *Mazzorno* and *Toi*, were united into one vast projecting cape, the most advanced point of which is now 32,000 or 33,000 *metres* * beyond the meridian of Adria; so that in the course of two hundred years, the mouths or

* From 19 miles 7 furlongs and 15 yards, to 20 miles 4 furlongs and 9 yards, English measure.—*Transl.*

delta of the Po have gained about 14,000 *metres* * upon the sea.

From all these facts, of which I have given a brief enumeration, the following results are clearly established.

First,—That at some ancient period, the precise date of which cannot be now ascertained, the waves of the Adriatic washed the walls of Adria.

Secondly,—That in the twelfth century, before a passage had been opened for the waters of the Po at *Ficarolo*, on its left or northern bank, the shore had been already removed to the distance of nine or ten thousand *metres* † from Adria.

Thirdly,—That the extremities of the promontories formed by the two principal branches of the Po, before the excavation of the *Taglio di Porto Viro*, had extended by the year 1600, or in four hundred years, to a medium distance of

* Or 15,366 yards.—*Transl.*
† Equal to 9,842 or 10,936 yards—*Transl.*

18,500 *metres* * beyond Adria; giving, from the year 1200, an average yearly increase of the alluvial land of 25 *metres.* †

Fourthly,—That the extreme point of the present single promontory, formed by the alluvions of the existing branches, is advanced to between thirty-two and thirty-three thousand *metres* ‡ beyond Adria; whence the average yearly progress is about seventy *metres* § during the last two hundred years, being greatly more rapid in proportion than in former times.

---

* Or 20,231 yards.—*Transl.*

† Exactly 27 yards 1 foot and 1-4th of an inch English.—*Transl.*

‡ Already stated at from 19¼ to 20½ miles; or more precisely, from 34.995 yards 1 foot 8 inches, to 36,089 yards 10 inches English measure.—*Transl.*

§ Equal to 76 yards 1 foot 7 inches and 9-10ths..—*Transl.*

APPENDIX;

CONTAINING

MINERALOGICAL NOTES,

AND AN ACCOUNT OF

CUVIER'S GEOLOGICAL DISCOVERIES.

NOTES.

Note A. § 4.

On the Subsidence of Strata.

M. Cuvier is of opinion, that all the older strata of which the crust of the earth is composed, were originally in an horizontal situation, and have been raised into their present highly-inclined position, by subsidences that have taken place over the whole surface of the earth.

It cannot be doubted, that subsidences, to a considerable extent, have taken place; yet we are not of opinion, that these have been so general as maintained by the illustrious author of this Essay We are rather inclined to believe, that the present inclined position of strata is in general their original one;—an opinion which is countenanced by the known mode of connection of strata, the phenomena of veins, particularly cotemporaneous veins, the crystalline nature of every species of older rock, and the great regularity in the *direction* of strata throughout the globe.

The transition and flœtz-rocks also are much more of a chemical or crystalline nature than has been generally imagined. Even sandstone, one of the most abundant of the flœtz-rocks, occasionally occurs in masses, many yards in extent, which individually have a tabular or stratified structure; but when viewed on the great scale, appear to be great massive distinct concretions. These massive concretions, with their subordinate tabular structures, if not carefully investigated, are apt to bewilder the mineralogist, and to force him to have recourse to a general system of subsidence or elevation of the strata, in order to explain the phenomena they exhibit.

Note B. § 7. p. 20 & 21.

On Primitive Rocks.

As the enumeration of primitive mountain rocks in the text is incomplete, we have judged it useful to give in this note a more full account of them. Primitive mountains, in general, form the highest and most rugged portions of the earth's surface, and extend in the form of chains of mountain-groups throughout the whole earth. These mountain-groups are generally highest in the middle, and lowest towards the sides and extremities; and the mountain-rocks of which they are composed, are so arranged, that in general the middle and highest portions of the group are composed of older rocks than the lateral and lower portions. As far as we know at present, granite is the oldest and first formed of all the primitive rocks. This rock is composed of felspar, quartz, and mica, and varies in its structure from coarse to very small granular. It sometimes alternates with beds of quartz and felspar, and is often traversed by cotemporaneous veins of granite, of quartz, and of felspar. The newer or upper portions of the formation contain cotemporaneous masses of porphyry, syenite, hornblende rock, limestone, &c. It frequently forms the highest, and at the same time the central part of mountain-groups. The next rock, in point of antiquity, or that which rests immediately upon the granite, is gneiss, which has a distinct slaty structure, is stratified, and, like granite, is composed of felspar, quartz, and mica. It alternates with the newer portions of the granite, and sometimes cotemporaneous veins of the one rock shoot into masses of the other. It contains subordinate formations of granite, porphyry, syenite, trap, quartz, limestone, and conglomerated gneiss. The next rock in the series is mica-slate, which rests upon the gneiss. It is composed of quartz and mica, and has a distinct slaty structure, and is stratified. It alternates with gneiss, and contains various subordinate formations, as granite, porphyry, syenite, trap, quartz, serpentine, limestone, and conglomerated mica-slate. It is often traversed by cotemporaneous veins, from the smallest discernible magnitude to many yards in width. The mica-slate is succeeded by clay-slate, which rests upon it, and sometimes alternates with it. It differs from mica-slate, gneiss, and granite, in its composition, being in general a simple rock; and in some instances principally composed of mica, in others to all appearance of felspar. Besides granite, porphyry, trap, syenite, limestone, serpentine, conglomerated clay-slate, * quartz, it also contains the following formations; flinty-slate, whet-slate, talk-slate, alum-slate, and drawing-slate. The calcareous rocks mentioned by Cuvier, as resting upon the slate, do not belong to this class; they are transition limestone, and contain, although rarely, testaceous petrifactions.

Note C. § 7. p. 21

Crystallised Marbles resting on shelly Strata.

M. Cuvier says, "the crystallised marbles never cover the shelly strata." This observation is not perfectly correct; for transition limestone, and certain magnesian floetz limestones, which are to be considered as crystallised marbles, contain testaceous petrifactions, and alternate with other strata that contain petrified shells.

Crystallised marble, or granular foliated limestone, occur, along with flœtz trap rocks in the coal formation, in different parts of Scotland, as upon the Lomonds, in Fife-shire, &c.

Note D. § 9. p. 25.

Salisbury Craigs.

The front of Salisbury craigs, near Edinburgh, affords a fine example of the natural chronometer, described in

* The primitive conglomerated rocks, mentioned above, as occurring in gneiss, mica-slate, and clay-slate, are sometimes named grey wacke.

the text. The acclivity is covered with loose masses that have fallen from the hill itself; and the quantity of debris is in proportion to the time which has elapsed since the waters of the ocean formerly covered the neighbouring country. If a vast period of time had elapsed since the surface of the earth had assumed its present aspect, it is evident, that long ere now the whole of this hill would have been enveloped in its own debris. We have here, then, a proof of the comparatively short period since the waters left the surface of the globe,—a period not exceeding a few thousand years.

Note E. § 10. p. 26.

On the Alluvial Land of the Danish Islands in the Baltic, and on the Coast of Sleswick.

In this section, Cuvier gives a clear and distinct account of several kinds of alluvial formations M. De Luc, in the first volume of his Geological Travels, describes the alluvial formations that cover and bound many of the islands in the Baltic and upon the coast of Denmark, and gives so interesting an account of the modes followed by the inhabitants in preserving these alluvial deposites, that we feel pleasure in communicating it to our readers.

" During my stay at Husum, I had the advantage of passing my evenings very agreeably and profitably at the house of M. Hartz, with his own family, and two Danish officers, Major Behmann, commandant at Husum, and Captain Baron de Barackow. The conversation often turned on the objects of my excursions, and particularly on the natural history of the *coasts* and of the *islands*; respecting which M. Hartz obligingly undertook to give me extracts from the chronicles of the country. This led us to speak of the Danish islands; and those officers giving me such descriptions of them as were very interesting to my object, I begged their permission to write down in their presence the principal circumstances which they communicated to me. These will form the first addition to my own observations; I shall afterwards proceed to the information which I obtained from M. Hartz.

"The two principal islands of the Danish Archipelago, those of Funen and Seeland (or Zeland), as well as some small islands in the Kattegate, namely, Lenoe, Anholt, and Samsoe, are hilly, and principally composed of *geest*;* and in these are found *gravel* and *blocks of granite*, and of other stones of that class, exactly in the same manner as in the country which I have lately described, and its *islands* in the North Sea. On the borders of the two first of these Danish islands, there are also *blocks* in the sea; but only in front of *abrupt* coasts, as is the case with the islands of Poel and Rugen, and along the coasts of the Baltic. The lands added to these islands of *geest* are in most part composed of the *sand* of the sea, the land-

* By *geest* is understood the alluvial matter which is spread over the surface both of the hilly and low country, and appears to have been formed the last time the waters of the ocean stood over the surface of the earth.—J.

waters there being very inconsiderable; and to the south of them have been formed several islands of the same nature, the chief of which are Laland and Falster, near Seeland. These, like the *marsch* islands in the North Sea, are sand-banks accumulated by the waves, and, when covered with grass, continuing to be farther raised by the sediments deposited between its blades. In the Baltic, where there are no sensible tides, such islands may be inhabited without dikes, as well as the extensions of the coasts; because, being raised to the highest level of that sea, while their declivity under water is very small, and being also more firm in their composition, the waves die away on their shores; and if, in any extraordinary case, the sea rises over them, it leaves on them fresh deposits, which increase their heights. These soils are all perfectly *horizontal*, like those added to the coasts of the continent.

" Some of these islands approach, entirely or in part, to the nature of that of Rugen. This island of Seeland, on that side which is called Hedding, has a promontory composed of strata of *chalk* with its flints. The island of Moen, (or Mona,) on the south of the latter, has a similar promontory near Maglebye and Mandemark; and the island of Bornholm, the easternmost of those belonging to Denmark, contains strata of *coal*, covered by others of *sandstone*. Phenomena like these, evident symptoms of the most violent catastrophes at the bottom of the ancient sea, proceeding, as I think I have clearly shewn, from the subsidence and angular motions of large masses of strata, which must have forced out the interior fluids with the utmost impetuosity, it is not surprising that so many fragments of the lowermost strata are found dispersed over this great theatre of ruins.

" I now proceed to the details which I received from M. Hartz; beginning by a specific designation of the *islands* dependent on the province of Sleswigh, such as they are at present, belonging to the three classes already defined. To commence from the north; Fanoe, Rom, Sylt, and Amrom, were originally *islands* of the same nature as the neighbouring continent, but have been since extended by *marsches*.* The soil of these islands, with its gravel and blocks of primordial stones, was at first barren, as the *geest* is naturally every where; but is become fertile by manure, of which there has been no deficiency, since those grounds have been surrounded with *marsch*, where the cattle are kept in stables during the winter. In the island of Sylt, there are spaces consisting of *moor;* but its head of land, which extends on the south as far as Mornum, is composed entirely of *marsch* and is bordered with *dunes* towards the open sea, because, the sediments of the rivers not reaching any farther, the *sea-sand* impelled against it by the waves remains pure, and is thus raised by the winds in hillocks

* By *marsch* is understood the new land added to the coasts since the last retiring of the water of the globe from the surface of the earth, and is formed by the sediments of rivers, mixed more or less with sand from the bottom of the sea.—J.

on the shore. The shallow bottom of the sea, between this island and that of Fora, is of *geest*: at low water, it may be passed over on foot; and there are found on it gravel and blocks of *granite*. But on the same side of Fora there is a great extent of *marsch*, beginning from St Laurencius. Among the islands consisting entirely of *marsch* and surrounded with dikes, the most considerable are Pellworm and Nord Strand; and among the Halligs, or those inhabited without dikes, the Chief are Olant, Nord-marsh, Langne, Groode, and Hooge.

" Such are the islands on this coast, in their present state, now rendered permanent by the degree of perfection at which the art of dike-making is arrived. But, in former times, though the *original* land was never attacked by the *sea*, which, by adding to it *new lands*, soon formed a barrier against its own encroachments, the latter, and the *islands* composed of the same materials, were subject to great and sudden changes, very fatal to those who were engaged to settle on them by the richness of their soil, comparatively with the continental. The inhabitants, who continued to multiply on them during several generations, were taught, indeed, by experience, that they might at last be invaded by the element which was incessantly threatening them; but having as yet no knowledge of natural causes, they blindly consider those that endangered them as supernatural, and for a long time used no precautions for their own security. They were ignorant of the dreadful effects of a certain association of circumstances, rare indeed, but, when occuring, absolutely destructive of these *marsches.* This association consists of an extraordinary elevation of the level of the North Sea, from the long continuance of certain winds in the Atlantic, with a violent storm occurring during the tides of the new or full moon; for then the sea rises above the level of all the *marsches;* and before they were secured against such attacks, the waves rolling over them, and tearing away the grass which had bound their surface, they were reduced to the sate of mere banks of sand and mud, whence they had been drawn by the long course of ordinary causes. Such were the dreadful accidents to which the first settlers on these lands were exposed; but no sooner were they over, than ordinary causes began again to act; the sand-banks rose; their surface was covered with grass; the coast was thus extended, and new islands were formed; time effaced the impression of past misfortunes; and those among the inhabitants of these dangerous soils, who had been able to save themselves on the coast, ventured to return to settle on them again, and had time to multiply, before the recurrence of the same catastrophes.

" This has been the general course of events on all the coasts of the North Sea, and particularly on those of the countries of Sleswigh and Holstein. It is thus that the origin and progress of the *art of dikes* will supply us with a very interesting *chronometer* in the history of the continent and of man, particularly exemplified in this part of the globe. A Lutheran clergyman, settled in the island of Nord Strand, having collected all the particulars of this history which the documents of the country could afford, published it in 1668, in a German work, entitled *The North Frisian Chronicle.* It was chiefly from this work, and from the *Chronicle of Dankwerth*, that M. Hartz extracted the information which he gave to me, accompanied by two maps, copied for me, by one of his sons, from those of Johannes Mayerus, a mathematician; they bear the title of *Frisia Cimbrica;* one of them respecting the state of the *islands* and of the *coast*, in 1240, as it may be traced in the chronicles, and the other, as it was in 1651.

"According to these documents, the first inhabitants of the *marsches* were *Frisii* or *Frisians*, designated also under the names of *Cimbri* and *Sikambri*: the latter name M. Hartz conjectures, might come from the ancient German words *Seekampfers*, i. e. *Sea-warriors;* the *Frisians* being very warlike. These people appear to have had the same origin with those, who, at a rather earlier period, took possession of the *marsches* of Ost-Frise, (East-Friesland,) and of that Friesland which forms one of the United Provinces; but this common origin is very obscure. Even at the present day, the inhabitants of the *marsches*, from near Husum to Tondern, or Tunder to the north, though themselves unacquainted with it, speak a language which the other inhabitants of the country do not understand, and which is supposed to be Frisian. It is the same at a village in the peninsula of Bremen, by which I have had occasion to pass.

" The *Sicambri* or *North Frisians*, are traced back to some centuries before the Christian era. At the commencement of that era, they were atacked by Frotho King of Denmark, and lost a battle, under their king Vicho, near the river Hever. Four centuries afterwards, they joined the troops of Hengist and Horsa. In the year 692, their king Radebot resided in the island of Heiligeland. Charles Martel subdued them in 732; and some time afterwards, they joined Charlemagne against Gottric, King of Denmark. These are some of the circumstances of the history of this Frisian colony, recorded in the chronicles of which I have spoken; but the history here interesting to us is that of the lands whereon they settled.

" It appears that these people did not arrive here in one body, but successively, in the course of many years: they spread themselves over various parts of the coasts of the north Sea, and even a considerable way up the borders of the Weser and the Elbe; according to documents which I have mentioned in my *Lettres sur l'Histoire de la Terre et de l'Homme*. These new settlers found large *marsches*, formed, as well in the wide mouths of those rivers as along the coasts, and around the original islands of *geest*; especially that of Heiligeland, the most distant from the coast, and opposite the mouth of the Eyder. Of this island, which is steep towards the south, the original mass consists of strata of *sandstone*; and at that time its *marsch* extended almost to Eyderstede: there were *marsches* likewise around all the other original islands; besides very large islands of pure *marsch* in the intervals of the former.

"All these lands were desert at the arrival of the Frisians; and the parts on which they established their first habitations, to take care of their breeds of horses and cattle feeding on the *marsches*, were the original eminences of the islands; on that of Heiligeland they built a temple to their great goddess Phoseta, or Fosta. When they became too numerous to confine themselves to the heights, their herds being also greatly multiplied, they ventured to begin inhabiting the *marsches*; but afterwards, some great inundations having shewn them the dangers of that situation, they adopted the practice followed by those who had settled on the *marsches* of the province of Groningen, and still continued in the Halligs; that of raising artificial mounts called *werfs*, on which they built their houses, and whither they could, upon occasion, withdraw their herds; and it likewise appears that, in the winter, they assembled in greater numbers on the spots originally the highest, in the islands, as well as on some parts of the coasts.

"Things continued in this state for several centuries; during which period, it is probable that the inhabitants of these lands were often, by various catastrophes, disturbed in the enjoyment of them, though not discouraged. But in 516, by which time these people were become very numerous, more than 600 of them perished by one of the concurrences of fatal circumstances already defined.

It was then that they undertook the astonishing enterprize of enclosing these lands. They dug ditches around all the *marsches*, heaping up on their exterior edge the earth which was taken out; and thus they opposed to the sea, dikes of eight feet in height. After this, comprehending that nothing could contribute more to the safety of their dwellings, than to remove the sea to a greater distance, they undertook, with that view, to exclude it from the intervals between the islands, by uniting, as far as should be possible, those islands with each other. I will describe the process by which they effected this, after I shall have recalled to attention some circumstances leading to it.

" From all that I have already said of the *fore-lands*, and of the manner in which they are increased, it may be understood, that the common effects of the *waves* and of the *tides* is to bring materials from the bottom of the sea towards the coasts; and that the process continues in every state of the sea. The land winds produce no *waves* on the coasts, which can carry back to the bottom of the sea what has been brought thence by the winds blowing against the shore; and as for the *tides*, it may have been already comprehended, (and shall soon be proved,) that the *ebb* carries back but very little of what has been brought by the *flood*. So that, but for some extraordinary circumstances, the materials continually impelled towards the shore, which first form islands, would at last unite against the coast in a continuous soil. The rare events, productive of great catastrophes, do not carry back these materials towards the bottom of the sea; they only, as it has been said before, ravage the surface, diminishing the heights, and destroying the effect of vegetation. These then were the effects, against which it was necessary to guard.

"I now come to the plan of uniting the *islands*, formed by these early inhabitants. They availed themselves for that purpose of all such parts of the sand-banks, as lay in the intervals between the large islands, and were beginning to produce grass. These, when surrounded with dikes, are what are called *Hoogs*; and their effects are to break the waves, thus diminishing their action against the dikes of the large islands, and at the same time to determine the accumulation of the mud in the intervals between those islands. In this manner a large *marsch* island, named Everschop, was already, in 987, united to Eyderstede by the point on which Poppenbull is situated; and in 995, the union of the same *marsches* was effected by another point, namely that of Tetenbull. Lastly, in the year 1000, Eyderstede received a new increase by the course of the Hever, prolonged between the sand-banks, being fixed by a dike; but the whole still remained an *island*. This is an example, of the manner in which the *marsch islands* were united by the *hoogs*; and the chronicle of the country says, that by these labours the islands were so considerably enlarged in size, and the intervals between them so much raised, that at low water it was possible to pass on foot from one to the other. The extent of these *marsches* was so great on the coast of Sleswigh alone, that they were divided into three provinces, two of which comprehended the *islands*, and the third comprised the *marsches* contiguous to the coast; and the same works were carried on upon the *marsches* of the coast of Holstein.

" But the grounds thus gained from the *sand-banks* were very insecure; these people, though they had inhabited them more than ten centuries, had not yet understood the possibility of that combination of fatal circumstances above described, against which their *dikes* formed but a very feeble rampart; the North Sea, by the extraordinary elevations of its level, being much more formidable in this respect than the ocean, where the changes of absolute level are much less considerable. I shall give an abridged account of the particulars extracted by M. Hartz from the chronicle of Dankwerth, relative to the great catastrophes which these *marsches* successively underwent, previously to the time when experience led to the means necessary for their security.

" In 1075, the island of Nord Strand, then contiguous to the coast, particularly experienced the effect of that unusual combination of destructive causes; the sea passing over its dike, and forming within it large excavations like lakes. In 1114 and 1158, considerable parts of Eyderstede were carried away; and in 1204, the part called Sudhever in the *marsch* of Uthholm was destroyed. All these catastrophes were fatal to many of the *marsch* settlers; but in 1216, the sea having risen so high that its waves passed over Nord Strand, Eyderstede, and Ditmarsch, near 10,000 of their inhabitants perished. Again, in 1300, seven parishes in Nord Strand and Pell-worm were destroyed; and in 1338, Ditmarsch experienced a new catastrophe, which swept away a great part of it on the side next Eyderstede: the dike of the course of the Eyder between the sand-banks was demolished, and the tides have ever since preserved their course throughout that wide space. Lastly, in the year 1362, the isles of Fora and Sylt, then forming but one, were divided, and Nord Strand, then a *marsch* united to the coast, was separated from it.

" During a long time, the inhabitants who survived these catastrophes, and their successors, were so much discouraged, that they attempted nothing more than to surround with *dikes* like the former such spaces of their meadow-land as appeared the least exposed to these ravages, leaving the rest to its fate. But the common course of causes continually tending to extend and to raise the grassy parts of the sand-banks, and no extraordinary combination of circumstances having interrupted these natural operations, later generations, farther advanced in the arts, undertook to secure to themselves the possession of those new grounds. In 1525, they turned their attention to the indentations made, during the preceding catastrophes, in the borders of the *marsches*; the waves, confined in these narrow spaces, sometimes threatening to cut their way into the interior part. In the front of all the creeks of this kind they planted stakes, which they interlaced with osiers, leaving a certain space between the lines. The waves, thus broken, could no longer do injury to the *marsch;* and their sediments being deposited on both sides of this open fence, very solid *fore-lands* were there formed. In 1550, they raised the *dikes* considerably higher, employing wheelbarrows, the use of which was only then introduced. For this purpose, they much enlarged and deepened the interior canals, in order to obtain more earth, not merely to add to the height of the dikes, but to extend their base on the outer side. At last they began to cover these dikes with straw ropes; but this great preservative of dikes was at first ill managed; and the use of it was so slowly spread, that it was not adopted in Nord Strand and in Eyderstede, till about the years 1610 and 1612.

" Before that time, however, the safety of the extensive soil of the latter *marsch* had been provided for in a different manner. I have said above, that, when the isles of Everschop and Uthholm had been united to it, the whole together still formed but one large *island;* now, in this state, it was in as great danger on the side towards the continent, as on that open to the sea; because two small rivers, the Trene and the Nord Eyder, discharging themselves into the interval between it and the land, and by preserving their course to the sea, this interval was thus kept open to tempests, sometimes from the side of the Hever, sometimes from that of the Eyder; and the waves, beating against the *geest,* were thence repelled upon the *marsch.* The inhabitants, seeing that the expence of remedying these evils would be greater than they could afford, while at the same time it was indispensable to their safety, addressed themselves to their bishop and to their prefect, of whom they requested pecuniary assistance; and having obtained it, they first undertook the great enterprize of carrying the Trene and the Nord Eyder higher up into the Eyder; keeping their waters, however, still separate for a certain space, by a *dam* with a *sluice*, in order to form there a reservoir of fresh water; the tides ascending up the Eyder above Frederichstadt. They were thus enabled to carry on the extremities of the *dike* on both sides to join the *geest*; and the interval between the latter and the *marsch* was then soon filled up, there being only left, at their junction, the canal above described, which receives the waters of the *geest*, and, at low water, discharges them from both its extremities by sluices. At the same time, the islands of Pellworm and Nord Strand were united with each other by means of eight *hoogs*; and the *sandy marsches* of which I have spoken, contiguous to the *geest*, on the north of that of Husum, were enclosed with dikes.

"After the dikes had been thus elevated, and their surface rendered firm by the straw ropes, though the latter were not yet properly fixed, the inhabitants of the *marsches* for some time enjoyed repose; but on the 11th of October 1634, the sea, rising to an excessive height, carried away, during a great tempest, the *hoogs* which had produced the junction between Pellworm and Nord Strand, these having ever since continued distinct islands; it also violently attacked Ditmarsch; and its ravages extended over the whole coast, as far as the very extensive new lands of Jutland. Princes then came forward zealously to the relief of their subjects. In particular, Frederic III. duke of Sleswigh, seeing that the inhabitants of Nord Strand were deficient both in the talents and in the means necessary for the reparation and future security of that large island, and knowing that the art of dikes had made greater progress in Holland, because of the opulence of the country, addressed himself to the States General, requesting them to send him an engineer of dikes with workmen accustomed to repair them; and this was granted. The dikes of Nord Strand were then repaired in the most solid manner; and the Dutch engineer, seeing the fertility of its soil, advised his sons, upon his death-bed, to purchase lands and settle there, if the duke would grant them the free exercise of their religion; they being Jansenist catholicks, and the inhabitants of the island Lutherans. The duke agreed to this, on condition that they and their posterity should continue to superintend the works carried on upon the dikes; to which they engaged themselves. From that time, the art of dikes, and particularly that part of it which consists in covering them solidly with straw, has become common to all the *marsches;* and the Dutch families, which have contributed to this fortunate change, continue to inhabit the same island, and to enjoy the free exercise of their religion."

NOTE F. § 11. p. 29.

On the Sand Flood.

In different parts of Scotland, there are examples of the natural chronometer, mentioned in the text. One of the most striking examples I at present recollect of this phenomenon in foreign countries, is that described by M. De Luc's brother, in the *Mercure de France*, for September 1807, and of which we here insert a translation:—

" The *sands* of the Lybian desert," he says, " driven by the west winds, have left no lands capable of tillage on any parts of the western banks of the Nile not sheltered by mountains. The encroachment of these *sands* on soils which were formerly inhabited and cultivated is evidently seen. M. Denon informs us, in the account of his *Travels in Lower and Upper Egypt*, that summits of the *ruins* of ancient *cities* buried under these *sands* still appear externally; and that, but for a ridge of mountains called the *Lybian chain*, which borders the left bank of the Nile, and forms, in the parts where it rises, a barrier against the invasion of these *sands*, the shores of the river, on that side, would long since have ceased to be habitable. Nothing can be more melancholy," says this traveller, " than to walk over villages swallowed up by the sand of the desert, to trample underfoot their roofs, to strike against the summits of their minarets, to reflect that yonder were cultivated fields, that there grew trees, that here were even the dwellings of men, and that all has vanished.

"If then our *continents* were as *ancient* as has been pretended, no traces of the habitation of men would appear on any part of the western bank of the Nile, which is exposed to this scourge of the *sands* of the desert. The existence, therefore, of such monuments attests the successive progress of the encroachments of the sand; and these parts of the bank, formerly inhabited, will for ever remain arid and waste. Thus the great population of Egypt, announced by the vast and numerous ruins of its cities, was in great part due to a cause of fertility which no longer exists, and to which sufficient attention has not been given. The *sands* of the desert were formerly remote from Egypt; the *Oases*, or habitable spots still appearing in the midst of the sands, being the remains of the soils formerly extending the whole way to the Nile; but these *sands*, transported hither by the western winds, have overwhelmed and buried this extensive tract, and doomed to sterility a land which was once remarkable for its fruitfulness.

"It is therefore not solely to her revolutions and changes of sovereigns that Egypt owes the loss of her ancient splendour; it is also to her having been thus irrecoverably deprived of a tract of land, by which, before the *sands* of the desert had covered it and caused it to disappear, her wants had been abundantly supplied. Now, if we fix our attention on this fact, and reflect on the consequences which would have attended it if thousands, or only some hundreds of centuries had elapsed since our continents first existed above the level of the sea, does it not evidently appear that all the country on the west of the Nile would have been buried under this *sand* before the erection of the cities of ancient Egypt, how *remote* soever that period may be supposed; and that, in a country so long afflicted with sterility, no idea would even have been formed of constructing such vast and numerous edifices? When these cities indeed were built, another cause concurred in favouring their prosperity. The navigation of the Red Sea was not then attended with any danger on the coasts: all its ports, now nearly blocked up with *reefs* of *coral*, had a safe and easy access; the vessels laden with merchandize and provisions could enter them and depart without risk of being wrecked on these shoals, which have risen since that time, and are still increasing in extent.

" The defects of the present government of Egypt, and the discovery of the passage from Europe to India round the Cape of Good Hope, are therefore not the only causes of the present state of decline of this country. If the *sands* of the desert had not invaded the bordering lands on the west, if the work of the *sea polypi* in the Red sea had not rendered dangerous the access to its coasts and to its ports, and even filled up some of the latter, the population of Egypt and the adjacent countries, together with their product, would alone have sufficed to maintain them in a state of prosperity and abundance. But now, though the passage to India by the Cape of Good Hope should cease to exist, though the political advantages which Egypt enjoyed during the brilliant period of Thebes and Memphis should be re-established, she could never again attain the same degree of splendour.

"Thus the *reefs* of *coral* which had been raised in the Red Sea on the east of Egypt, and the *sands* of the desert which invade it on the west, concur in attesting this truth: That our continents are not of a more remote *antiquity* than has been assigned to them by the sacred historian in the book of Genesis, from the great era of the Deluge."

Note G. § 15. p. 33.
On Coral Islands.

Of all the genera of lithophytes, the madrepore is the most abundant. It occurs most frequently in tropical countries, and decreases in number and variety as we approach the poles. It encircles in prodigious rocks and vast reefs many of the basaltic and other rocky islands in the South Sea and Indian Ocean, and by its daily growth adds to their magnitude. The coasts of the islands in the West Indies, also those of the islands on the east coast of Africa, and the shores and shoals of the Red Sea, are encircled and incrusted with rocks of coral. Several different species of madrepore contribute to form these coral reefs; but by far the most abundant is the muricated madrepore, madrepra muricata of Linnæus. These lithophytic animals not only add to the magnitude of land already existing, but, as Cuvier remarks, they form whole islands. Dr Forster, in his Observations made during a Voyage round the World, gives the following account of the formation of these coral islands in the South Sea.

"All the low isles seem to me to be a production of the sea, or rather its inhabitants, the polype-like animals forming the lithophytes. These animalcules raise their habitation gradually from a small base, always spreading more and more, in proportion as the structure grows higher. The materials are a kind of lime mixed with some animal substance. I have seen these large structures in all stages, and of various extent. Near Turtle Island, we found, at a few miles distance, and to leeward of it, a considerable large circular reef, over which the sea broke every where, and no part of it was above water; it included a large deep lagoon. To the east and north-east of the Society Isles, are a great many isles, which, in some parts, are above water; in others, the elevated parts are connected by reefs, some of which are dry at low-water, and others are constantly under water. The elevated parts consist of a soil formed by a sand of shells and coral rocks, mixed with a light black mould, produced from putrified vegetables, and the dung of sea-fowls; and are commonly covered by cocoa-nut trees and other shrubs, and a few antiscorbutic plants. The lower parts have only a few shrubs, and the above plants; others still lower, are washed by the sea at high-water. All these isles are connected, and include a lagoon in the middle, which is full of the finest fish; and sometimes there is an opening, admitting a boat or canoe in the reef, but I never saw or heard of an opening that would admit a ship.

"The reef, or the first origin of these isles, is formed by the animalcules inhabiting the lithophytes. They raise their habitation within a little of the surface of the sea, which gradually throws shells, weeds, sand, small bits of corals, and other things, on the tops of these coral rocks, and at last fairly raises them above water; where the above things continue to be accumulated by the sea, till by a bird, or by the sea, a few seeds of plants, that commonly grow on the sea shore, are thrown up, and begin to vegetate; and by their annual decay and reproduction from seeds, create a little mould, yearly accumulated by the mixture with sand, increasing the dry spot on every side; till another sea happens to carry a cocoa-nut hither, which preserves its vegetative power a long time in the sea, and therefore will soon begin to grow on this soil, especially as it thrives equally in all kinds of soil; and thus may all these low isles have become covered with the finest cocoa-nut trees.

"The animalcules forming these reefs, want to shelter their habitation from the impetuosity of the winds, and the power and rage of the ocean; but as, within the tropics, the winds blow commonly from one quarter, they, by instinct, endeavour to stretch only a ledge, within which is a lagoon, which is certainly entirely screened against the power of both: this therefore might account for the method employed by the animalcules in building only narrow ledges of coral rocks, to secure in their middle a calm and sheltered place: and this seems to me to be the most probable cause of THE ORIGIN of all THE TROPICAL LOW ISLES, over the whole South Sea."

That excellent navigator, the late Captain Flinders, gives the following interesting account of the formation of Coral Islands, particularly of Half-way Island on the north coast of Terra Australis * :

" This little island, or rather the surrounding reef, which is three or four miles long, affords shelter from the south-east winds; and being at a moderate day's run from Murray's Isles, it forms a convenient anchorage for the night to a ship passing through Torres' Strait: I named it *Half-way Island.* It is scarcely more than a mile in circumference, but appears to be increasing both in elevation and extent. At no very distant period of time, it was one of those banks produced by the washing up of sand and broken coral, of which most reefs afford instances, and those of Torres' Strait a great many. These banks are in different stages of progress: some, like this, are become islands, but not yet habitable; some are above high-water mark, but destitute of vegetation; whilst others are overflowed with every returning tide.

" It seems to me, that when the animalcules which form the corals at the bottom of the ocean, cease to live, their structures adhere to each other, by virtue either of the glutinous remains within, or of some property in salt water; and the interstices being gradually filled up with

---

* Vol. II. p. 114, 115, 116.

sand and broken pieces of coral washed by the sea, which also adhere, a mass of rock is at length formed. Future races of these animalcules erect their habitations upon the rising bank, and die in their turn to increase, but principally to elevate, this monument of their wonderful labours. The care taken to work perpendicularly in the early stages, would mark a surprising instinct in these diminutive creatures. Their wall of coral for the most part, in situations where the winds are constant, being arrived at the surface, affords a shelter, to leeward of which their infant colonies may be safely sent forth; and to this their instinctive foresight it seems to be owing, that the windward side of a reef exposed to the open sea, is generally, if not always, the highest part, and rises almost perpendicular, sometimes from the depth of 200, and perhaps many more fathoms. To be constantly covered with water, seems necessary to the existence of the animalcules, for they do not work, except in holes upon the reef, beyond low-water mark; but the coral sand and other broken remnants thrown up by the sea, adhere to the rock, and form a solid mass with it, as high as the common tides reach. That elevation surpassed, the future remnants, being rarely covered, lose their adhesive property; and remaining in a loose state, form what is usually called a *key*, upon the top of the reef. The new bank is not long in being visited by sea birds; salt plants take root upon it, and a soil begins to be formed; a cocoa-nut, or the drupe of a pandanus, is thrown on shore; land birds visit it, and deposit the seeds of shrubs and trees; every high tide, and still more every gale, adds something to the bank; the form of an island is gradually assumed; and last of all comes man to take possession.

"Half-way Island is well advanced in the above progressive state; having been many years, probably some ages, above the reach of the highest spring tides, or the wash of the surf in the heaviest gales. I distinguished, however, in the rock which forms its basis, the sand, coral, and shells, formerly thrown up, in a more or less perfect state of cohesion. Small pieces of wood, pumice stone, and other extraneous bodies which chance had mixed with the calcareous substances when the cohesion began, were inclosed in the rock; and in some cases were still separable from it without much force. The upper part of the island is a mixture of the same substances in a loose state, with a little vegetable soil; and is covered with the *casuarina* and a variety of other trees and shrubs, which give food to parroquets, pigeons, and some other birds; to whose ancestors, it is probable, the island was originally indebted for this vegetation."

Note H. § 16. p. 35.

On the Diminution of the Waters of the Ocean.

That the water of the ocean has diminished, and is still diminishing, can scarcely be doubted; yet the rate of decrease since the period of the deluge has been so gradual, being now effected not by the conversion of the water into the earthy materials of which the globe is composed, but principally by the agency of animals, vegetables, and volcanoes, that, on a general view, it may be said to be nearly imperceptible. The facts mentioned by Celsius and others, in regard to the rapid diminution of the waters of the Baltic, have been much insisted on by some geologists, although they cannot correctly be employed in illustrating the supposed general diminution of the waters of the globe; because the Baltic is a nearly inclosed sea, receiving rivers of considerable magnitude. One of the best writers of the age, when discussing this topic, says—

"If we proceed further to the north, to the shores of the *Baltic* for instance, we have undoubted evidence of a *change of level* in the same direction as on our own shores. The level of the sea has been represented as lowering at so great a rate as *forty inches in a century*. Celsius observed, that several rocks which are now above the water, were not long ago sunken rocks, and dangerous to navigators; and he took particular notice of one, which in the year 1680, was on the surface of the water, and in the year 1731 was 20½ Swedish inches above it. From an inscription near *Aspo*, in the lake *Melar*, which communicates with the *Baltic*, engraved, as is supposed, about five centuries ago, the level of the sea appears to have sunk in that time no less than thirteen Swedish feet. All these facts, with many more which it is unnecessary to enumerate, make the *gradual depression*, not only of the *Baltic*, but of the whole *Northern Ocean*, a matter of certainty."—PLAYFAIR'S *Illustrations*, p. 445.

That indefatigable and accurate observer De Luc, has the following commentary on the preceding passage:—

"It would be unnecessary to mention even the two inconsiderable facts above, if the *depression* of the *level* of these *seas* were indeed *a matter of certainty*; for the best authenticated and the least equivocal monuments of their *change* would then abound along all their coasts. But proofs are every where found that such a *change* is *chimerical*: they may be seen in all the *vales* coming down to these *seas*, in which there is no perceptible impression of the action of any *waters* but those of the *land*, and no vestige, through their whole extent, of any permanent abode of those of the *sea*: and proofs to the same effect are equally visible, along the coasts of both these *seas*, in all the *new lands* which have been formed on them, and which, being perfectly *horizontal* from the point where their formation commenced, evidently shew that the *water* displaced by them has been constantly at the same *level*. Hence appears the necessity of multiplying, as I have done and shall continue to do, for the subversion of a prejudice of such ancient date, the examples of these peremptory proofs of its total want of foundation. The *rock* mentioned by Celsius had probably been observed by him at times when the *level* of the *sea* was different; its known differences much exceeding the quantity here specified. As for the inscription near Aspo, in a country abounding with *lakes* as much as that which I have above described, if we were acquainted with its terms, we should probably find it to be, like many which I have seen in various places along the course of the Oder and the Elbe, the monument of some extraordinary inundation of the land, from the sudden melting of the snows in the mountains, at a time when the water had been prevented from running off by an equally extraordinary rise of the level of the sea; of which the effects on low coasts may extend very far inland.

"By his conclusion, however, from these few facts, contrary to every thing observed on the coasts of this sea, Mr Playfair thinks himself authorized to maintain that *the gradual depression, not only of the Baltic, but of the whole northern ocean, is a matter of certainty:* afterwards he examines merely which of these two causes, the *subsidence* of the *sea* itself, or the *elevation* of the *land* around it, agrees the best with the phenomena; and he decides in favour of the latter, pointing out its accordance with the Huttonian theory."

Note I. (A.) § 23.

Werner's Views of the Natural History of Petrifactions.

From the observation in section 22, Cuvier does not appear to have known how much Werner has done for the advancement of the natural history of fossil organic remains. He did not rest satisfied with the developement of the mere mineralogical branch of the theory of the earth; on the contrary, early in life he began to investigate the relations of all the classes of fossil organic remains, being well convinced, that without an accurate and comprehensive knowledge of these interesting bodies, geological speculation would have excited but comparatively little notice. Many years ago he embodied all that was known of petrifactions into a regular system. He insisted on the necessity of every geognostical cabinet containing, besides complete series of rocks for illustrating the mineralogical relations of the globe, an extensive collection not only of shells, but also of the various productions of the class zoophyta, of plants, particularly of sea plants and ferns; and an examination of the remains of quadrupeds in the great limestone caves and alluvial soils of Germany, soon pointed out to him the necessity of attaching to the geognostical cabinet also one of comparative osteology. As his views in geognosy enlarged, he saw more and more the value of a close and deep study of petrifactions. He first made the highly important observation, that different formations can be discriminated by the petrifactions they contain. It was during the course of his geognostical investigations that he ascertained the general distribution of organic remains in the crust of the earth. He found that petrifactions appear first in transition rocks. These are but few in number, and of animals of the zoophytic or testaceous kinds. In the older flœtz rocks they are of more perfect species, as of fish and amphibious animals; and in the newest flœtz and alluvial rocks, of birds and quadrupeds, or animals of the most perfect kinds. He always maintained that no fossil remains of the human species had been found in flœtz rocks, or in any of the older alluvial formations; but was of opinion that such remains might be discovered in the very newest of the alluvial depositions. He also was led to believe, from his numerous observations, that sea plants were of more ancient origin than land plants. A careful study of the genera and species of petrifactions disclosed to him another important fact, viz. that the petrifactions contained in the oldest rocks are very different from any of the species of the present time; that the newer the formation, the more do the remains approach in form to the organic beings of the present creation; and that in the very newest formations, fossil remains of the present existing species occur. He also ascertained, that the petrifactions in the oldest rocks were much more mineralized than those in the newer rocks, and that in the newest rocks they were merely bleached or calcined. He found that some species of petrifactions were confined to particular beds; others were distributed throughout whole formations, and others seemed to occur in several different formations; the original species found in these formations appearing to have been so constituted as to live through a variety of changes which had destroyed hundreds of other species, which we find confined to particular beds.

Note I. (b) § 23.

On the Distribution of Petrifactions in the different classes of Rocks.

As an account of the distribution of fossil organic remains throughout the strata, of which the crust of the earth is composed, cannot fail to prove interesting, even to the general reader, we shall here give a very short sketch of what is known on the subject. Fossil organic remains, or petrifactions, have not hitherto been discovered in any of the primitive rocks; indeed it would appear that animals and vegetables were not called into existence until the period when the transition rocks began to be formed.

Hence it is, that petrifactions have not been met with in any rock older than those of the transition class.

TRANSITION ROCKS.

The principal transition rocks are greywacke, greywacke slate, clay slate, limestone, greenstone, amygdaloid, syenite, porphyry, and granite. All of them do not afford petrifactions, these bodies having been hitherto found only in limestone, greywacke, greywacke slate, and clay slate.

1. *Transition Limestone.*

Fossil corallitic bodies, such as madreporites, tubiporites, and milleporites, of different species, abound in many varieties of this limestone. It is in general difficult to determine the species of these genera, owing to their being much intermixed with each other, and with the matter of the limestone. On a general view, they certainly approach very near, in external characters, to those corals we at present meet with in a living state in the tropical regions of the globe. Intermixed with these corals, or in separate strata, we find various species of orthoceratites, lituites, ammonites, belemnites, nautilites, lenticulites, chamites, terebratulites, anomites, and patellites.

2. *Greywacke.*

This is a rock, including in a basis of quartzy clay slate, variously shaped, masses of clay slate, greywacke slate, flinty slate, and sometimes also masses and grains of felspar, and scales of mica. It very rarely contains petrifactions. Hence in many extensive tracts of country where it predominates, not a single fossil organic remain is to be seen. The animal petrifactions which have been discovered in this rock are ammonites, and madreporites, of the same species as those met with in clay slate, and greywacke slate; also solenites, mytulites, and tellinites, large orthoceratites, and, according to some naturalists, remains of animals of the serpent kind. The vegetable petrifactions are fruits, stems and leaves of palm-like vegetables, and parts of reeds.

3. *Clay Slate.*

It rarely contains petrifactions; and the only kinds hitherto met with in it appear to be ammonites and trilobites.

4. *Greywacke Slate.*

This rock seldom contains petrifactions. Where it borders on the clay slate, it contains the same kinds of ammonites as occur in that rock, and in the vicinity of grey-wacke and transition limestone, we observe in it orthoceratites, corallites, and fossil remains of reeds and marine plants. The orthoceratites gracilis of Blumenbach, the Molossus of Montfort, and also the coralliolites orthoceratoides, which are found in this rock, seem to belong to those remarkable corals that form a kind of connecting link between shells and corals. Particular beds of a siliceous and ferruginous nature, subordinate to the greywacke slate, abound more in petrifactions. They contain principally some species of madreporites; also screw-stones, (schraubensteine), which appear to be derived from the coralliolites epithonius, and whole families of terebratulites, with a few species of turbinites and striped chamites.

It appears from the preceding statement, that in general the different species of transition rocks contain similar petrifactions, and that they are principally distinguished by the number of corals and orthoceratites imbedded in them.

FLŒTZ ROCKS.

Fossil organic remains are much more abundant, and more varied in the rocks of this than of the preceding class. We shall enumerate the rocks of this class according to their relative antiquity, and begin with the lowest or first formed member of the series, which is named

I. *First Sandstone, or Old Red Sandstone.*

This rock is characterised by its colour, composition, imbedded minerals, strata with which it is associated, the veins that traverse it, and its position in regard to the other rocks of which the crust of the earth is composed. Interposed between it and the first flœtz limestone, and even sometimes alternating with it, there occurs an extensive coal formation, which contains, besides the coal, beds of sandstone, slate clay, bituminous shale, clay ironstone, limestone, claystone, and various trap rocks.* The red sandstone contains but few petrifactions,

---

* The coal formations in Northumberland and Durham, described by Dr Thomson in his Annals of Philosophy for 1814, appear to belong to that above described; and his magnesian limestone may be a member of the first or oldest flœtz limestone formation. The coal formation in the neighbourhood of Edinburgh probably occupies the same place in this series.

and these are principally of trunks or branches of trees, some of which appear to resemble those of the tropical regions. In the sandstone which is associated with the coal, and also in the slate clay with which it alternates, there frequently occur remains of common and of arborescent ferns, gigantic reeds, palms, and leaves of a tree which resembles the *casuarina*, and which was long considered as an *equisetum*. In the limestone, slate clay, &c. of the coal fields in this country, many petrifactions occur, such as orthoceratites, ammonites, nautilites, serpulites, patellites, helicites, turbites, buccinites, trochites, mytulites, cardites, anomites, pectinites, echinites, entrochites, and milleporites. Bones and teeth of fishes have been also found in the coal formation.

II. *First Flœtz Limestone.*

This limestone rests immediately on the first sandstone formation. It is divided into the following members: 1. Alpine limestone. 2. Bituminous marl slate. 3. Zechstein. 4. The coal subordinate to the formation in general.

1. *Alpine Limestone.**

This is the most highly crystallised limestone of the series. It is principally characterised by the ammonites and lenticulites it contains. In it we also meet with single coralliolites, encrinites, terebratulites, ostracites, buccinites, chamites, echinites, belemnites, and gryphites.

* It would appear from the observations of Mohs, that this alpine limestone belongs to the transition class.

2. *Bituminous Marl Slate.*

This remarkable limestone is very widely distributed, and often contains abundance of petrified fishes, which are in general most numerous in those places where the rock occurs in basin shaped strata. Many attempts have been made to determine the genera and species of these animals, but hitherto with but little success. It would appear that the greater number are fresh water species, and a few marine species. But the most remarkable fossil organic remain hitherto found in this limestone, is that of an animal of the genus monitor, of the class amphibia, of which Cuvier has given an interesting account in his great work on Fossil Organic Remains.

Petrifactions of vegetables rarely occur in this limetone; we sometimes meet with branches of plants analogous to the *lycopodium*, and more rarely fragments of *ferns*, and of plants allied to the genus *phalaris*.

Amongst these fresh water productions, we meet with various fossil remains of marine animals, such as gryphites, pentacrinites, trilobites, and corallophites.

3. *Zechstein.*

This rock, in some of its characters, resembles the alpine limestone, but does not contain so many petrifactions. Ammonites occur in it; and pentacrinites fasciculosus, and whole families of gryphites aculeatus. It contains more rarely the gryphites rugosus, terebratulites alatus, terebratulites lacunosus, and probably also the terebratulites striatissimus. T. obliquus, and T. variabilis. It affords nearly the same species of milleporites and coralliolites as are found in the bituminous marl slate. It is worthy of remark, that nearly all the petrifactions found in this formation are much broken.

4. *Coal.*

Beds of coal occur in the zechstein, and also, according to some mineralogists, in the alpine limestone, accompanied with slate clay, bituminous slate, and other rocks, all of which frequently contain petrifactions of bivalve shells, and impressions of plants. The shells resemble those met with in the alpine limestone, and also in the Jura limestone; and the vegetable impressions are of lycopodiums and ferns, resembling those found in the old coal formation. But, besides these, we observe remains of plants of the palm tribe, some of which resemble the carica papaya, a native of Senegal.

III. *Second and Third Sandstone Formations.*

The second sandstone rests in the first limestone and gypsum, but the position of the third sandstone has not been accurately ascertained. The following are some of the petrifactions mentioned by authors as occurring in them.

Encrinites trochitiferus. Schlottheim. Brunswick.
Dentalites striatus. Schlottheim. Mecklenburg.
Trochilites scheuchzeri. St Gallen.
Turbinites torquatus. Knorr. Neufschatel.
regensbergensis. Knorr. Regenberg, near Blankenburg.

Turbinites australis. Schlottheim. France.
Muricites volutinus. Bourg. T. 34. F. 223. St Gallen.
nisus. Bourg. T. 34. F. 226. St Gallen.
assimilis. Bourg. T. 24. F. 228. St Gallen.
Bullites reticulatus. Bourg. T. 37. F. 247. St Gallen.
senilis. Bourg. T. 37. F. 250. St Gallen.
Pectinites punctatus. Volkm. Siles. subterr. T. 23. F. 3.
radiatus. Id. T. 32. F. 6.
reticulatus. Id. T. 33. F. 1.
longicolli. Id. T. 33. F. 9.
anomalus. Id. T. 34. F. 13.
gigas. Knorr. P. II. 1. T. B. F. 1. 2. Ortenberg.
polonicus. Schlottheim. Wieliczka.
Chamites transversim punctatus. Volkm. Siles. subterr. T. 33. F. 7.
Ostracites labiatus. Knorr. P. II. 1. T. B. II. *b*** Fy. 2. Pirna.
Anomites paradoxus. Scheuchz. F. 96.
Pinnites diluvianus. Knorr. P. II. 1. T. D. X. F. 1. 2 Pirna.
Gryphites rugosus. Knorr. P. II. 1. T. B. 1. d. F. 7. Wieliczka.
Musculites sablonatus. Bourg. T. 23. F. 142. 143.
rugosus. Knorr. P. II. 1. T. B. vi. F. 3. Silesia.
Tellinites musculitiformis. Knorr. P. II. 1. T. B. II. St Gallen.
margaritaceus. Schlottheim. Mecklenburg.

IV. *Second Flœtz Limestone, or Jura Limestone.*

This formation, which rests on the rocks of the second sandstone formation, and is remarkable for the abundance and variety of petrifactions it contains, includes beds of coal, marl, sand-stone, stink-stone, and probably also of gypsum. The petrifactions occur principally in the beds of marl, sand-stone and stink-stone, and more sparingly in the other strata.

The following are the genera of petrifactions that have been met with in it :—serpulites, asterialites, encrinites, echinites, orthoceratites, belemnites, ammonites, nautilites, lenticulites, helicites, trochilites, buccinites, patellites, chamites, buccardites, donacites, venulites, ostracites, terebratulites, anomites, gryphites, musculites, and coralliolites. Some varieties contain petrified fishes of various genera and species, and also fossil amphibious animals. The vegetable petrifactions that occur in this formation are of stems and leaves, as those of the populus and rhamnus, and of flowers, as the ranunculus.

V. *Third Flœtz, or Shell Limestone.*

This formation is newer than either the second limestone or sandstone; and the following list contains the names of several of the petrifactions found in it.

Asteriatites eremita. Schlottheim. Gotha.
Encrinites trochitiferus. Blumenb. Abbild. F. 60.
Pentacrinites Gottingensis. Heimberg, near Gottingen,

Pentacrinites Britannicus. Blum. Abbild. T. 70. F. *a. b.* Dorsetshire. *

Echinites ruralis. Schlottheim. Tonna.

Dentalites obsoletus. Schlottheim. Tonna.

Bitubulites problematicus. Bl. Abb. T. II. F. 9.

Belemnites paxillosus. Schlottheim. Heimberg, near Gottingen.

Ammonites nodosus. Mus. Tessin. T. 4. F. 3. Thuringia.

franconicus. Knorr. P. II. 1. A. 2. F. 1. Koburg.

margaritatus. Montf. Fol. 90. Antwerp.

amaltheus. Knorr. P. II. 1. T. A. II. F 3. France.

planulites. Montf. F. 78.

dubius. Bourg. T. 39. F. 163.

spatosus. List. Anim. Angl. T. 6. F. 3. Gottingen.

pusillus. Schlottheim. Heimberg.

papiraceus. Schlottheim. Heimberg.

æneus. Bourg. T. 40. F. 266.

Nautilites pseudopompilus. Schlottheim. Weimar.

rusticus. Schlottheim. Heimberg, near Gottingen.

Helicites girans. Oryct. nor. T. III. F. 29.

planorbiformis. Schlottheim. Near Arensberg Thurengia.

pseudopomarius. Knorr. T. B. vi. *a.* F. 10. Quedlenburg.

Trochilites speciosus. Oryct. Nor. T. vii. F. 20.

nodosus. Schlottheim. Heimberg.

---

* Does this really belong to the shell limestone ?

Trochilites umbilicatus. Schlottheim. Heimberg.
læ vis. Schlottheim. Heimberg.
acutus. Schlottheim. Heimberg.
Neritites spiratus. Shlottheim. Arensburg.
gryphus. Schlottheim. Minden.
Turbinites strombiformis. Naturf. 1. S. 1. T. III. F. 3. Palatinate.
communis. Schlottheim.
socialis. Schlottheim. Wissbaden.
approximatus. Schlottheim. Heimberg.
Strombites Jenensis. Know. P. II. 1. T. C. vi. F. 7. Jena.
canaliculatus. Schlottheim. Heimberg.
Buccinites annulatus. Schlottheim. Halberstadt.
gregarius. Schlottheim. Heimberg.
Porcellanites Seelandicus. Schlottheim. Zeeland.
Patellites Vinariensis. Naturf. 5. St. T. III. F. 4. Weimar.
Discites æquilateralis. Schlottheim. Tonna.
Chamites lævis Bourg. T. 31. F. 120.
auritus. List Anim. Angl. T. 9. F. 51.
striatus. Bourg. T. 25. F. 154.
sulcatus. List Anim. Angl. T. 9. F. 54.
Pectinites subreticulatus. Schlottheim. Teutleben.
Buccardites cordicalis. Oryct. Nor. T. 7. F. 29.
cardissæformis. Schlottheim. Heimberg.
Donacites clausus. Schlottheim. Tonna.
Venulites trigonatus. Schlottheim. Tonna.
Ostracites sulcatus. Blumenb. Spec. Arch. Tel. T. 1. F. 3.
plicatus. Knorr. P. II, 1. T. D. *i*. F. 1—4.
pusillus. Oryct. Nor. T. viii. F. 8.
pyramidans. Oryct. Nor. T. iv. F. 1.

Ostracites spondyloides. Schlottheim. Tonna.
Terebratulites communis. Knorr. P. II. 1. T. B. iv. F. 2.
giganteus. Blumenb. Abb. T. i. F. 4. Osnabruck.
regularis. Oryct. Nor. T. v. F. 23.
oblongus. Oryct. Nor. T. v. F. 24.
squamiger. Oryct. Nor. T. v, F. 19.
artifex. Knorr. P. II. 1. T. B. iv. F. 7. 8.
sustarcinatus. Oryct. Nor. T. vii. F. 35.
subhistericus. Oryct. Nor. T. vii. F. 37.
parasiticus. Schlottheim. Tonna.
fragilis. Schlottheim. Herda.
bicanaliculatus. Schlottheim. Tonna.
Trigonellites pes anseris. Knorr. P. II. 1. T. B. II. *b.* F. 8. Thuringia.
communis. Knorr. P. II. 1. T. B. II. *b.*
simplex. Schlottheim. Sachsenberg.
Anomites obsoletus. Schlottheim. Lohberg.
Solennites annulatus. Oryct. Nor. T. iv. F. 12. 13. Winkelheid.
Gryphites.
Ratisbonensis. Knorr. P. II. 1. T. D. III. *c.* F. 1—3.
suillus. Schlottheim. Heimberg.
lævis. Schlottheim. Heimberg.
Musculites gibbosus. Oryct. Nor. T. vii. F. 25.
comprimatus. Oryct. Nor. T. vii. F. 23.
mytiloides. Oryct. Nor. iv. F. 2.
Pholadites caudatus. Halberstadt.
Mytilites sociatus. Thuringia.

Mytilites costatus. Lohberg, near Tonna.
Tellinites paganus. Oryct. Nor. T. vii. F. 26. 27.
comprimatus. Sachsenburg.
minutus. Schlottheim. Sachsenburg.
Balanites porosus. Blumenb. Abb. T. i. F. 1. Near Osnabruck.
parasiticus. Lohberg. Tonna.
Trilobites cornigerus. Schlottheim. Near Reval.

Fossil remains of fishes, and, it is said, also of birds, have been found in this formation.

VI. *Chalk Formation.*

This, which is one of the newest of the flœtz limestones, contains many different petrifactions, as will appear from the following enumeration.

Serpulites contortuplicatus. Mont. P. II. p. 25. Petersberg.
peniformis. Schlottheim. Petersberg.
exuviatus. Schlottheim. Island Rugen.
Osteriatites siderolites. Mont. P. I. p. 150. Petersberg.
Asteriatites spinosus. Schlottheim. Petersberg.
pentagonatus. Schlottheim. Petersberg.
Echinites poundianus. Schlottheim. Kent.
Echinites varians. Bourg. T. li. F. 337—339.
anomalus. List. Anim. Angl. T. vii. F. 25.
melitensis. List. Anim. Angl. T. xxvii.
cordiformis. List. Anim. Angl. T. vii. F. 28.
Breynianus. Breyn. Opuscl. T. iv. F. 1. 2.

Echinites fenestratus. Knorr. T. E. 7. *a*. T. iii.
canaliculatus. Knorr. P. II. 1. T. E. iv. F. 1. 2.
ursinus. Knorr. P. II. 1. T. E. 1. *a*. F. 4.
hexagonatus. Knorr. P. II. 1. T. E. V. F. 12.
cruciatus. Knorr. Suppl. T. ix. *d*. F. 3.
sideralis. Naturf. 9 St. T. iv. F. 7. Petersberg.
echinometrites. Bourg. T. liii. F. 361.
Dentalites minutus. Schlottheim. Island Moen.
Orthoceratites gigas. Knorr. Suppl. T. xii. F. 1—5.
Telebois annulatus. Montf. P. I. p. 366. Island of Gothland.
Baculites vertebralis. Montf. P. I. p. 343.
Belemnites reticulatus. Montf. P. I. p. 379. St Catherine.
pyrgopolon mosæ. Montf. P. I. p. 394.
mucronatus. Breyn. opuscl. Tabula Belemnit. T. 1. *a*. 2. *b*. Faujas.
paxillosus. Montf. P. I. p. 352.
lanceolatus. Breyn. Tab. Bel. F. 7. *a*.
Ammonites mammillatus. Naturf. 1. St. T. II. F. 3.
elipsolites funatus. Montf. P. I. p. 86. St Catherine.
Nautilites pseudopompilius. Fauj. Peterberg. T. xxi. F. 1.
puppis. Fauj. T. xxv. F. 9. Petersberg.
pulcher. Fauj. T. xx. F. 3. Petersberg.
Srombites globulatus. Knorr. P. II. 1. T. C. vii.
Buccinites Belgicus. Petersberg.
Muricites turrilitis costatus. Montf. P. I. 118. Rouen.
Volutites coniformis. Knorr. P. II. 1. T. C. ii.* F. 6. 7.
Patellites acutus. Fauj. T. xxv. F. 1. Petersberg.

Patellites mitratus. Knorr P. II. ii. T. N. F. 3. Mecklenburg.
melitensis. Knorr. P. II. I. T. B. 1. *c*. F. 5. 6. Suppl. T. v. *c*. F. 6.
regularis. Fauj. T. T. xxiii. F. 2. Petersberg.
irregularis. Fauj. T. xxiii. F. 3. Petersberg.
Ostracites mysticus. Fauj. T. xxvi. F. 5. Petersberg.
ungulatus. Knorr. P. II. 1. T. D. vii. F. 5. 6. Petersberg.
orista urogalli. Knorr. P. II. 1. T. D. vii. F. 3. 6.
laurifolium. Knorr. P. II. 1. T. D. vii. F. 1. 2.
plicatissimus. Naturf. 9 St. T. iv. F. 6. *a*—*b*. Kent.
approximatus. Fauj. T. xxiii. F. 5. Petersberg.
crista meleagris. Fauj. T. xxiii. F. 6. Petersberg.
haliotiformis. Fauj. T. xxiii. F. 4. Petersberg.
mactroides. Schlottheim. Champagne.
Terebratulites communis. Fauj. T. xxvi. F. 5. Petersberg.
scaphula. Fauj. T. xxvi. F. 8.
chrysalis. Fauj. T. xxvi. F. 7. & 9.
varians. Fauj. T. xxvi. F. 1.
microscopicus. Fauj. T. xxvi. F. 2.
limbatus. Fauj. T. xxvi. F. 4.
chitoniformis. Fauj. T. xxvi. F. 6.
peltatus. Fauj. T. xxvi. F. 11.
plicatellus. Fauj. T. xxvi. F. 10.
vermicularis. Fauj. T. xxvi. F. 12.
pectiniformis. Fauj. T. xxvii. F. 5.

Terebratulites tenuissimus. Fauj. T. xxvii. F. 7.
concavus. Fauj. T. xxvii. F. 6.
papillatus. Fauj. T. xxvii. F. 8.
gracilis. Schlottheim. Kent.
Pinnites cretaceus. Fauj. T. xxii. F. 1. & 3.
Gryphites politus. Schlottheim. Island Moen.
Tellinites asserculatus. Knorr. Suppl. T. v. *c.* F. 2. Mecklenburg.

Besides these petrifactions, the following are enumerated by authors as occurring in chalk: spondylites, pectinites, chamites, teeth and bones of fish, also fish much mutilated, tortoises, crabs, alcyonites, madreporites, spongites, and encrinites. *

VII. *Flœtz Trap Rocks.*

These rocks occur in several of the flœtz formations already mentioned, either as subordinate beds, or in mountain masses. In the red sandstone formations they occur in beds, veins, and mountain masses, and appear in single hills, as Salisbury Craig near Edinburgh, or in ranges of hills, as the Pentlands and Ochils, also near Edinburgh. The only rock of the series which contains petrifactions is the trap-tuff, which includes a few vegetable impressions.

* I enumerate in this list the petrifactions discovered by Faujas St Fond in the Petersberg, near Mæstricht, as it is the opinion of some naturalists that it belongs to the chalk formation.

Flœtz trap rocks also occur in the flœtz limestone formation, either in beds or mountain masses; and sometimes we meet with whole ranges of such hills belonging to the flœtz limestone. I do not know that petrifactions have ever been found in the trap of these formations.

The Coal Formation, which forms a great tract of country on both sides of the frith of Forth, contains beds and veins of flœtz trap rocks. The only trap rock of this series which contains petrifactions is the trap-tuff, and it very rarely presents a few vegetable impressions.

VIII. *Newest Flœtz Trap.*

The newest flœtz trap formation of Werner, which is of a very late date, contains very few petrifactions. From the short account now given, it appears, that the flœtz trap rocks, in whatever situation they occur, contain very few organic remains. *

IX. *Newest Flœtz Formations.*

Over the chalk rests a series of calcareous and siliceous formations, which, in general, abound in petrifactions. They appear to have been deposited from the water of lakes or inland seas, some of which are conjectured to have been alternately filled with fresh and salt water; and hence,

* This is the formation considered by many geologists as entirely of volcanic origin.

in a general view, are of a more local nature than those which have been deposited from the waters of the ocean. The newest members of the series are of so loose a texture, the fossil organic remains they contain so nearly resemble those that now inhabit the earth, and they are so nearly related to the alluvial formations which are daily forming, that it is often extremely difficult, nay even sometimes impossible, to determine whether they belong to the alluvial or newest flœtz formation. There appears to be a gradation or transition from the one into the other. The petrifactions they contain are of zoophytes, shells, fishes, and amphibious animals; and fossil remains of birds and quadrupeds here for the first time appear enclosed in strata. The country around Paris, that of the Isle of Wight, and other districts in the south of England, as particularly described in Note K (B), belong to these newer formations.

X. *Alluvial Formations.*

The mineral substances included under this class are considered to be of newer formation than any of the flœtz rocks; and the following are the most frequent and abundant of these, viz. gravel, sand, clay, loam, marl, calc-tuff, calc-sinter, brown coal, and peat.

Petrifactions frequently occur distributed through these deposites, either in a regular or irregular manner, and are sometimes whole, sometimes more or less broken, but angular, or are so much rounded as to shew that they have suffered by attrition. Several different alluvial formations may be pointed out, which are characterised by the organic remains they contain. Thus, one formation found in this neighbourhood contains shells of the common oyster, common muscle, patella vulgaris, buccinum undatum and lapillus, nerita littoralis, and turbo littoreus, all of which are still inhabitants of the Frith of Forth. Another contains bones of ruminating animals, as those of the horse, ox, and stag, but differing from those of the living species; and in a third, which contains much marl and many fresh water shells, there occur the bones of several extinct species of the elephant, rhinoceros, hippopotamus, and also of the Irish elk, which is no longer a native of this country.

From the preceding short sketch it appears, that the most simple animals are those we first meet with in a mineralised state; that these are succeeded by others more perfect, and which are contained in newer formations; and that the most perfect, as quadrupeds, occur only in the newest formations. But we naturally inquire, have no remains of the human species been hitherto discovered in any of the formations? Judging from the arrangement already mentioned, we would naturally expect to meet with remains of man in the newest of the formations. In the writings of ancient authors, there are descriptions of anthropolithi. In the year 1577, Fel. Plater, Professor of Anatomy at Basil, described several fossil bones of the elephant found at Lucerne, as those of a giant at least nineteen feet high. The Lucernese were so perfectly satisfied with this discovery, that they caused a painting be made of the giant as he must have appeared when alive, assumed two such giants as the supporters of the city arms, and had the painting hung in their public hall. The Landvoigt Engel, not satisfied with this account of these remains, maintained that our planet, before the creation of the present race of men, was inhabited by the fallen angels, and that these bones were parts of the skeletons of some of those miserable beings. Scheuchzer published an engraving and description of a fossil human skeleton, which proved to be a gigantic species of salamander or proteus. Spallanzani describes a hill of fossil human bones in the island of Cerigo; but this also is an error, as has been satisfactorily shewn by Blumenbach. Lately, however, a fossil human skeleton has been imported into this country from Guadaloupe by Sir Alexander Cochrane. It is imbedded in a block of calcareous stone, composed of particles of limestone and coral, and which, like the aggregations of shells found on the limestone coasts in some parts of this country, has acquired a great degree of hardness. It is therefore an instance of a fossil human petrifaction in an alluvial formation. The engraving here given is copied from the Philosophical Transactions of the Royal Society of London; and the following description of the fossil remains it exhibits is that of Mr Konig, which has been drawn up with great care.

" The situation of the skeleton in the block was so superficial, that its presence in the rock on the coast had probably been indicated by the projection of some of the more elevated parts of the left fore-arm.

" The operation of laying the bones open to view, and

of reducing the superfluous length of the block at its extremities, being performed with all the care which its excessive hardness and the relative softness of the bones required, the skeleton exhibited itself in the manner represented in the annexed drawing (Pl. III.), with which my friend Mr Alexander has been so good as to illustrate this description.

"The skull is wanting; a circumstance which is the more to be regretted, as this characteristic part might possibly have thrown some light on the subject under consideration, or would, at least, have settled the question, whether the skeleton is that of a Carib, who used to give the frontal bone of the head a particular shape by compression, which had the effect of depressing the upper and protruding the lower edge of the orbits, so as to make the direction of their opening nearly upwards, or horizontal, instead of vertical. *

"The vertebræ of the neck were lost with the head. The bones of the thorax bear all the marks of considerable concussion, and are completely dislocated. The seven true ribs of the left side, though their heads are not in connexion with the vertebræ, are complete; but only three of the false ribs are observable. On the right side only fragments of these bones are seen; but the upper part of the seven true ribs of this side are found on the

* See the excellent figures in Blumenbach's Decades.

left, and might at first sight be taken for the termination of the left ribs; as may be seen in the drawing. The right ribs must therefore have been violently broken and carried over to the left side, where, if this mode of viewing the subject be correct, the sternum must likewise lie concealed below the termination of the ribs. The small bone dependent above the upper ribs of the left side, appears to be the right clavicle. The right os humeri is lost; of the left nothing remains except the condyles in connexion with the fore-arm, which is in the state of pronation; the radius of this side exists nearly in its full length, while of the ulna the lower part only remains, which is considerably pushed upwards. Of the two bones of the right fore-arm, the inferior terminations are seen. Both the rows of the bones of the wrists are lost, but the whole metacarpus of the left hand is displayed, together with part of the bones of the fingers: the first joint of the fore-finger rests on the upper ridge of the os pubis; the two others, detached from their metacarpal bones, are propelled downwards, and situated at the inner side of the femur, and below the foramen magnum ischii of this side. Vestiges of three of the fingers of the right hand are likewise visible, considerably below the lower portion of the fore-arm, and close to the upper extremity of the femur. The vertebræ may be traced along the whole length of the column, but are in no part of it well defined. Of the os sacrum, the superior portion only is distinct: it is disunited from the last vertebra and the ilium, and driven upwards. The left os ilium is nearly complete, but shattered, and one of the fragments depressed below the level of the rest; the ossa pubis, though well defined, are gradually lost in the mass of the stone. On the right side, the os innominatum is completely shattered, and the fragments are sunk; but towards the acetabulum, part of its internal cellular structure is discernible.

"The thigh bones and the bones of the leg of the right side are in good preservation, but being considerably turned outwards, the fibula lies buried in the stone, and is not seen. The lower part of the femur of this side is indicated only by a bony outline, and appears to have been distended by the compact limestone that fills the cavities both of the bones of the leg and thigh, and to the expansion of which these bones probably owe their present shattered condition. The lower end of the left thigh bone appears to have been broken and lost in the operation of detaching the block; the two bones of the leg, however, on this side, are nearly complete; the tibia was split almost the whole of its length a little below the external edge, and the fissure being filled up with limestone, now presents itself as a dark coloured straight line. The portion of the stone which contained part of the bones of the tarsus and metatarsus, was unfortunately broken; but the separate fragments are preserved.

"The whole of the bones, when first laid bare, had a mouldering appearance, and the hard surrounding stone could not be detached without frequently injuring their surface; but after an exposure for some days to the air, they acquired a considerable degree of hardness. Sir

H. Davy, who subjected a small portion of them to chemical analysis, found that they contained part of their animal matter, and all their phosphate of lime."

Note K. (a) § 28. p. 103.

Cuvier's Geological Discoveries.

As the Essay on the Theory of the Earth does not contain a full account of the numerous geological discoveries and observations of Professor Cuvier, we shall lay before our readers a condensed view of the most important of these, drawn up from his great work on the Fossil Remains of Quadrupeds.

The first volume of this invaluable work contains, besides the present Essay on the Theory of the Earth, a masterly delineation of the geognostic structure of the country around Paris. It would appear, from the description there given, that the fundamental rock or basis of the district is *chalk*. This chalk is covered with *plastic clay*, and what is termed *coarse marine limestone*. The limestone abounds in marine petrifactions, and is associated with a kind of siliceous limestone, which contains the well-known mineral in the arts, used as a millstone, and named *buhrstone*. Over this limestone rests a remarkable formation of *gypsum*. It alternates with beds of marl, containing menilite, and beds of clay, with imbedded lenticular crystals of gypsum. The gypsum contains remains of extinct quadrupeds, birds, amphibious animals, fishes, and shells, all of which are said to be land or fresh-water species; hence it is denominated *a fresh-*

water formation. Above this gypsum lie beds of *marl* and *sandstone* that contain marine shells, thus affording another marine formation. These rocks are covered with beds of *millstone*, limestone and flint, both of which contain petrifactions of fresh-water shells; hence this association is named the *second fresh-water formation.* The uppermost formation is of an *alluvial* nature. It is composed of variously-coloured sand, marl, clay, or a mixture of these substances, impregnated with carbon, which gives the mixture a brown or black colour. It contains rolled stones of different kinds, but is most particularly characterised by containing the remains of large organic bodies. It is in this formation that we find great trunks of trees, bones of elephants, also of oxen, rein-deer, and other mammalia. From the intermixture of fresh and salt-water organic productions in these formations, we may suppose that both these fluids must have contributed each their part in their formation. According to Cuvier, and Brongniart, who assisted him in examining these formations we have just enumerated, there appears to have been an alternate flux and reflux of salt and fresh water over the country around Paris, and from which these rocks were deposited. The descriptions are accompanied with a geognostical map and sections of the strata in different parts of the country around Paris. The whole is executed according to the method of the Wernerian geognosy.

The Second Volume contains admirable osteological descriptions of several of the larger species of quadrupeds, as the rhinoceros, hippopotamus, tapir, and elephant. These are given with the view of elucidating the fossil species which are here described. Eleven fossil species are investigated, viz. one rhinoceros, two hippopotami, two tapirs, one elephant, and five mastodons. The species are described in the following order.

1. *Rhinoceros.*—Three species of this genus are at present known to naturalists, as inhabitants of different parts of the world. These are the two-horned rhinoceros of Africa, the one-horned rhinoceros of Asia, and the rhinoceros of the island of Sumatra. Only one fossil species has hitherto been discovered, which differs from the three living species, not only in structure, but in geographical distribution. It was first noticed in the time of Grew, and the bones he mentions were dug out of alluvial soil near Canterbury. Since that period similar remains have been found in many places of Germany, France, and Italy. In Siberia, not only single bones and skulls, but the whole animal, with the flesh and skin, have been discovered.

2. *Hippopotamus.*—Only one species of this genus is at present known to live on the surface of the earth. It is an inhabitant of Africa, and, according to Marsden, also of Asia, for he mentions it as one of the animals of the island of Sumatra. M. Cuvier is inclined to call in question the accuracy of this statement of Marsden's, and to conjecture that he may have confounded the Succotyro of Newhoff with the hippopotamus. Mr Marsden, in the

new edition of his excellent description of Sumatra, still enumerates the hippopotamus amongst the Sumatrian animals, but appears to have misunderstood Cuvier, when he says that he accuses him of confounding the hippopotamus with the *dugong*.* Two fossil species have been ascertained by Cuvier. The one, which is the largest, is so very nearly allied to the species at present living on the surface of the earth, that it is difficult to determine whether or not it is not the same. Its fossil remains have been found in alluvial soil in France and Italy. The second fossil species, and the smallest, the animal not being larger than a hog, is well characterised, and is entirely different from any of the existing species of quadrupeds.

---

* "Hippopotamus, *Küda-ayer*. The existence of this quadruped in the island of Sumatra having been questioned by M. Cuvier, and not having myself actually seen it, I think it necessary to state, that the immediate authority upon which I included it in the list of animals found there, was a drawing made by *M. Whalfeldt*, an officer employed in a survey of the coast, who had met with it at the mouth of one of the southern rivers, and transmitted the sketch along with his report to the government, of which I was then secretary. Of its general resemblance to that well-known animal there could be no doubt. M. Cuvier suspects that I may have mistaken it for the animal called by naturalists the *dugong*, and vulgarly the sea-cow, which will be hereafter mentioned; and it would indeed be a grievous error, to mistake for a beast with four legs, a fish with two pectoral fins, serving the purposes of feet; but, independently of the authority I have stated, the *küda-ayer*, or river-horse, is familiarly known to the natives, as is also the duyong (from which Malayan word the dugong of naturalists has been corrupted); and I have only to add, that in a register given by the Philosophical Society of Batavia, in the first volume of their Transactions, for 1799, appears the article, '*conda aijeer*, rivier paard, hippopotamus,' amongst the animals of Java."—MARSDEN's *History of Sumatra*. 3d edit. p. 116, 117.

3. *Tapir.*—The tapir is an animal peculiar to the new world, and has hitherto been found only in South America. Yet two fossil species of this genus have been discovered in Europe. The one is named the small, the other the gigantic tapir, and both have been found in different parts of France, Germany, and Italy.

4. *Elephant.*—Of this genus two species are at present known as inhabitants of the earth. The one, which is confined to Africa, is named the African elephant; the other, which is a native of Asia, is named the Asiatic elephant. Only one fossil species has hitherto been discovered. It is the *mammoth* of the Russians. It differs from both the existing species, but agrees more nearly with the Asiatic than the African species.* Its bones have been found in many different parts of this island; as in the alluvial soil around London, in the county of Northampton, at Gloucester, at Trenton, near Stafford, near Harwich, at Norwich, in the island of Sheppey, in the river Medway, in Salisbury Plain, and in Flintshire in Wales; and similar remains have been dug up in the north of Ireland. Bones of this animal have been dug up in Sweden, and Cuvier conjectures that the bones of supposed giants, mentioned by the celebrated Bishop Pontoppidan as having been found in Norway, are remains of the fossil elephant. Torfæus mentions a head and tooth of this animal dug up in the island of Iceland.

* These three species are well distinguished by the appearance of the surface of the grinding teeth, as is shewn in *plate second*.

In Russia in Europe, Poland, Germany, France, Holland, and Hungary, teeth and bones of this species of elephant have been found in abundance. Humboldt found teeth of this animal in North and South America. But it is in Asiatic Russia that they occur in greatest abundance. Pallas says, that from the Don or the Tanais to Tchutskoinoss, there is scarcely a river the bank of which does not afford remains of the mammoth; and these are frequently imbedded in, or covered with, alluvial soil, containing marine productions. The bones are generally dispersed, seldom occurring in complete skeletons, and still more rarely do we find the fleshy part of the animal reserved. One of the most interesting instances on record of the preservation of the carcase of this animal, is given by M. Cuvier in the following relation: *

"In the year 1799, a Tungusian fisherman observed a strange shapeless mass projecting from an ice-bank, near the mouth of a river in the north of Siberia, the nature of which he did not understand, and which was so high in the bank as to be beyond his reach. He next year observed the same object, which was then rather more disengaged from among the ice, but was still unable to conceive what it was. Towards the end of the following summer, 1801, he could distinctly see that it was the frozen carcase of an enormous animal, the entire

* This singular discovery is given by Professor Cuvier, as taken from a Report in the Supplement to the *Journal du Nord*, No. xxx. by M. Adams, adjunct member of the Academy of St Petersburgh.

flank of which and one of its tusks had become disengaged from the ice. In consequence of the ice beginning to melt earlier and to a greater degree than usual in 1803, the fifth year of this discovery, the enormous carcase became entirely disengaged, and fell down from the ice-crag on a sand-bank forming part of the coast of the Arctic Ocean. In the month of March of that year, the Tungusian carried away the two tusks, which he sold for the value of fifty rubles; and at this time a drawing was made of the animal, of which I possess a copy.

"Two years afterwards, or in 1806, Mr Adams went to examine this animal, which still remained on the sand-bank where it had fallen from the ice, but its body was then greatly mutilated. The *Jukuts* of the neighbourhood had taken away considerable quantities of its flesh to feed their dogs; and the wild animals, particularly the white bears, had also feasted on the carcase; yet the skeleton remained quite entire, except that one of the fore-legs was gone. The entire spine, the pelvis, one shoulder-blade, and three legs, were still held together by their ligaments and by some remains of the skin; and the other shoulder-blade was found at a short distance. The head remained, covered by the dried skin, and the pupil of the eyes was still distinguishable. The brain also remained within the skull, but a good deal shrunk and dried up; and one of the ears was in excellent preservation, still retaining a tuft of strong bristly hair. The upper-lip was a good deal eaten away, and the under-lip was entirely gone, so that the teeth were distinctly seen.

The animal was a male, and had a long mane on its neck.

" The skin was extremely thick and heavy, and as much of it remained as required the exertions of ten men to carry away, which they did with considerable difficulty. More than thirty pounds weight of the hair and bristles of this animal were gathered from the wet sand-bank, having been trampled into the mud by the white bears, while devouring the carcase. Some of the hair was presented to our Museum of Natural History by M. Targe, censor in the Lyceum of Charlemagne. It consists of three distinct kinds. One of these is stiff black bristles, a foot or more in length; another is thinner bristles, or coarse flexible hair, of a reddish-brown colour; and the third is a coarse reddish-brown wool, which grew among the roots of the long hair. These afford an undeniable proof that this animal had belonged to a race of elephants inhabiting a cold region, with which we are now unacquainted, and by no means fitted to dwell in the torrid zone. It is also evident that this enormous animal must have been frozen up by the ice at the moment of its death.

" Mr Adams, who bestowed the utmost care in collecting all the parts of the skeleton of this animal, proposes to publish an exact account of its osteology, which must be an exceedingly valuable present to the philosophical world. In the mean time, from the drawing I have now before me, I have every reason to believe that the sockets of the teeth of this northern elephant have the same proportional lengths with those of other fossil elephants, of which the entire skulls have been found in other places." *

5. *Mastodon.*—This is entirely a fossil genus, no living species having hitherto been discovered in any part of the world. It is more nearly allied to the elephant than to any other animal of the present creation; it appears to have been a herbivorous animal; and the largest species, the *great mastodon* of Cuvier, was equal in size to the elephant. Five species are described.

1. *Great Mastodon.*—This species has been hitherto found in greatest abundance in North America, near the river Ohio, and remains of it have been dug up in Siberia. It has been frequently confounded with the mammoth or fossil elephant, and in North America it is named mammoth. In *plate 2d* we have given an engraving of one of the grinding teeth of this animal.

2. *Mastodon with narrow Grinders.*—The fossil remains of this species have been dug up at Simorre and many other places in Europe, and also in America.

---

* It is worthy of remark, that although fossil bones of the elephant were described as such in the middle of the 16th century by Aldrovandus, it was not until two centuries afterwards that this opinion was credited. In the intermediate time they were described as lusus naturæ, bones of giants, *skeletons of fallen angels*, remains of marine animals, or of colossal baboons.

3. *Little Mastodon with small Grinders.*—This species is much less than the preceding, and was found in Saxony and Montabusard.

4. *Mastodon of the Cordilleras.*—This species was discovered in South America by Humboldt. Its grinders are square, and it appears to have equalled in size the great mastodon.

5. *Humboldien Mastodon.*—This, which is the smallest species of the genus, was found in America by Humboldt.

All the fossil species of quadrupeds we have just enumerated have been found in the alluvial soil which covers the bottoms of vallies, or is spread over the surface of plains. All of them are strangers to the climate where these bones now rest. The five species of mastodons alone may be considered as forming a distinct and hitherto unknown genus, nearly allied to that of the elephant. All the others belong to genera still existing in the torrid zone. Three of these genera, viz. the rhinoceros, hippopotamus, and elephant, occur only in the old world; the fourth, the tapir, exists only in the new world. But the fossil species have not the same geographical distribution: It is in the old world that we dig up the bones of the tapir, and some remains of the elephant have been discovered in the new world. The fossil species included under the known genera differ sensibly from the present species, and are certainly not mere varieties. Of all the eleven fossil species, the large hippopotamus is the only one which we cannot say with certainty does not belong to the present living species of that genus. The small hippopotamus and gigantic tapir are unquestionably new species; there is scarcely a doubt of the fossil rhinoceros being a distinct species; and although the fossil elephant and the little tapir are not so well marked as new species, yet, as Cuvier remarks, there are reasons sufficient to convince the experienced anatomist of their being different from any of the present existing species. These different fossil bones are found almost every where in beds of nearly the same kind; they are often promiscuously mixed with bones of animals resembling the species of the present time. These beds are generally alluvial, either sandy or marly, and always near the earth's surface. It is therefore probable that these bones have been enveloped by the last, or one of the last, catastrophes to which our earth has been subjected. In many places they are accompanied with accumulations of marine animal remains, and in other places the sand and marl which cover them contain only fresh water shells. We have no authentic account of their having been found covered with flœtz, or other solid strata containing marine animals, and therefore cannot affirm that they were for a long time covered with a tranquil sea. The catastrophe, then, which has covered them, appears to have been a transient marine inundation. This inundation does not appear to have reached to the high mountains, because the formation in which these remains are found does not occur there, and these bones are not found in the high vallies, if we except a few in the warmer parts of America. The bones are neither rolled nor in skeletons, but dispersed, and in part broken or fractured. They have not therefore been brought there from a distance by an inundation, but have been found by it in the places where it has covered them, as might be expected, if the animals to which they belonged had dwelt in these places, and had there successively died. Hence it appears, that before this catastrophe these animals lived in the countries where we now find their bones: It is this inundation which has destroyed them; and as we do not find them elsewhere, the species must have been annihilated. It would thus appear, that the northern parts of the globe formerly nourished species belonging to the elephant, hippopotamus, rhinoceros, tapir, and mastodon tribes; and all of these, with exception of the mastodon, which is entirely a fossil genus, have species living, but only in the torrid zone. Nevertheless there is nothing to countenance the belief, that the species of the torrid zone have descended from the ancient animals of the north, which have been gradually or suddenly transported toward the equator. They are not the same; and we may see, by the examination of the most ancient mummies, as those of the ibis, that no established fact authorises the belief of changes so great as those which must be assumed for such a transformation, especially in wild animals. Nor are there any decisive proofs of the temperature of northern climates having changed since this epoch. The fossil species do not differ less from the living, than certain northern animals differ from their co-genera of the south;—the *isatis* of Siberia, for example, (*canis lagopus*) from the *chacal* of

India and of Africa (*canis aureus*). They therefore ought to have belonged to much colder climates.

The Third Volume contains particular descriptions of the various fossil bones that occur in the gypsum quarries around Paris, and of the fossil remains of birds, amphibious animals, and fishes associated with these.

The investigation of the fossil remains in the gypsum of Paris, has enabled Cuvier to add two entirely new genera of quadrupeds to the system of animals. These he entitles *Palæotherium* and *Anoplotherium*, and arranges them in the system in the following manner.

Class. Mammalia.
Ordo. Pachyderm.

Genus I. Palæotherium, * (Pone Tapirum et ante Rhinocerotem et Equum ponendum.

Dentes 44. *Primores utrinque* 6.
Laniarii 4, *acuminati paulo longiores, tecti.*
Molares 28, *utrinque* 7. *Superiores quadrati; inferiores bilunati.*
Nasus productior, flexilis.
Palmæ et plantæ tradactylæ.

---

* Palæotherium signifies ancient large animal, or beast.

1. P. Magnum. *Statura Equi.*
2. P. Medium. *Statura Suis; pedibus strictis, subelongatis.*
3. P. Crassum. *Statura Suis; pedibus latis, brevioribus.*
4. P. Curtum. *Pedibus ecurtatis patulis.*
5. P. Minus. *Statura Ovis; pedibus strictis, digitis lateralibus minoribus.*

Besides these five species found in the gypsum quarries around Paris, remains of others have been discovered in other parts of France, either imbedded in the *fresh water limestone*, or in alluvial soil. Cuvier enumerates and describes the following species.

6. P. Giganteum. *Statura Rhinocerotis.*
7. P. Tapiroides. *Statura Bovis; molarium inferiorum colliculis fere rectis, transversis.*
8. P. Buxovillanum. *Statura Suis; molaribus inferioribus extus sub gibbosis.*
9. P. Aurelianensi. *Statura Suis; molarium inferiorum angulo intermedio bicorni.*
10. P. Occitanicum. *Statura Ovis; molarium inferiorum angulo intermedio bicorni.*

Genus II. Anoplotherium.* (Inter Rhinocerotem aut Equum, ab una, et Hippopotamum, Suem et Camelum, ab altera parte ponendum.)

Dentes 44, *serie continua.*

Primores utrinque 6.

Laniarii primoribus similes, ceteris non longiores.

Molares 28, *utrinque* 7. *Anteriores compressi. Posteriores superiores quadrati. Inferiores bilunati.*

Palmæ et plantæ didactylæ, ossibus metacarpi et metatarsi discretis; digitis accessoriis in quibusdam.

1. A. Commune. *Digito accessorio duplo breviori, in palmis tantum; cauda corporis longitudine crassissima.*
Magnitudo Asini aut Equi minoris.
Habitus elongatus et depressus Lutræ.
Verisimiliter natatorius.

2. A. Secundarium. *Similis præcedenti, sed statura Suis. E tibia et molaribus aliquot cognitum.*

3. A. Medium. *Pedibus elongatis, digitis accessoriis nullis.*
Magnitudo et habitus elegans Gazellæ.

---

* Anoplotherium signifies beast without weapons; thus referring to its distinguishing character, its want of canine teeth.

4. A. Minus. *Digito accessorio utrinque, in palmis et plantis, intermedios fere æquante.*
Magnitude et habitus Leporis.

5. A. Minimum. *Statura paviæ Cobayæ, e maxilla tantum cognitum.*
Habitatio omnium, olim, in regione ubi nunc Lutetia Parisiorum.

The first eight Memoirs of this volume are occupied with descriptions of the two new genera already described, viz. the Palæotherium and Anoplotherium.

The ninth Memoir contains an account of the remains of carnivorous animals found in the gypsum quarries around Paris. Three species are mentioned; one nearly resembles the common fox; the second appears to belong either to an unknown species of canis, or to a genus intermediate between canis and viverra; the third is allied to the ichneumon, but is double the size.

In the tenth Memoir we have the description of a nearly entire skeleton of a quadruped of the genus Didelphis, found in the Paris gypsum. It does not belong to any of the present existing species, and is therefore said to be extinct. M. Cuvier here remarks, that as all the species of this genus are natives of America, it is evident that the hypothesis advanced by some naturalists, of all these fossil organic remains having been flooded from Asia to northern countries, is erroneous.

The eleventh Memoir treats of the fossil remains of birds found near Paris.

Cuvier, contrary to the opinion of many naturalists, proves that fossil remains of birds do exist. He does not, however, pretend to have ascertained with accuracy the species. Thus some of the bones he conjectures to belong to the genus pelicanus; others to the order grallæ; some to the genus sturnus (starling); and bones of the tern and quail tribes appear intermixed with the others.

The twelfth Memoir contains descriptions of the bones of reptiles and fishes found in the gypsum quarries around Paris.

Hitherto but few remains of these animals have been discovered in the vicinity of Paris. Cuvier describes only three species of amphibia, and five of fish. The amphibious animals are of the tortoise and crocodile tribes. The tortoises are different from any of the present existing species, and have been inhabitants of fresh water; the crocodile is also different from any of the living species known to naturalists. Of the fossil fish, the first belongs to a new genus nearly allied to *amia*, and is probably a fresh water fish. The second is nearly allied to two fresh water genera, viz. the *mormyrus* of La Cepide, natives of the river Nile, and the *pœcilia* of Bloch, natives of the fresh waters of Carolina. The third appears to be a species of sparus, different from any of the present species. The fourth and fifth very dubious.

It appears from the details in this volume, that all the reptiles and fishes, and we may add also, the shells found in the gypsum around Paris, prove that the beds containing the remains of the palæotherium, anoplotherium, and other unknown quadrupeds, have not been formed in the water of the ocean, but have been precipitated from fresh water.

The Fourth Volume, which is also extremely interesting, contains descriptions and histories of the fossil remains of bisulcated or ruminating animals, of horses, of animals of the hog tribe, of glires or gnawing animals, of carnivorous and other clawed animals, and of oviparous quadrupeds, or animals of the class amphibia of Linnæus.

The first Memoir is confined to the history and description of animals of the order *bisulca*, found in alluvial strata.

Cuvier remarks, that the osteological study of the fossil remains of the animals of this division is very difficult. The general resemblance to each other of these animals is so great, that the several genera can only be characterised by parts, such as horns, which, from their frequently varying with age, sex, and climate, must, in their fossil and mutilated states, be very uncertain guides.

He first describes the fossil remains of the species of the deer tribe, and in the following order.

1. *Fossil Elk of Ireland.*—This is the most celebrated of all the fossil ruminating animals. It is most certainly a different species from any of those that at present live on the earth's surface, and may therefore be considered as extinct. It was first found in Ireland, where it generally occurs in shell marl and in peat-bogs. It has also been found in superficial alluvial soil in England, Germany, and France.

In *plate 1st*, we have given a drawing of the head and horns of this animal. It was dug out of a marl-pit at Dardistoun, near Drogheda, in Ireland. Dr Molyneux, in the Philosophical Transactions, informs us that its dimensions were as follow :—

		Feet.	Inches.
From the extreme tip of each horn, .	a. b.	10	10
From the tip of the right horn to its root, ...	c. d.	5	2
From the tip of one of the inner branches to the tip of the opposite branch, .	e. f.	3	7½
The length of one of the palms, within the branches, ...	g. h.	2	6
The breadth of the palm, within the branches, ...	i. k.	1	10½
The length of the right brow antler, .	d. l.	1	2
The beam of each horn, at some distance from the head, in diameter, ...	m.	0	$2\frac{1}{10}$
in circumference, . . .		0	8

		Feet.	Inches.
The beam of each horn, at its root, in circumference, ...	d.	0	11
The length of the head, from the back of the skull to the extremity of the upper jaw, ...	n. o.	2	0
Breadth of the skull, ...	p. q.	1	0

2. *Fossil Deer of Scania.*—This species of fossil deer was found in a peat-moss in Scania. It appears, from the description of the horns, to be an extinct, or at least an unknown species.

3. *Fossil Deer of Somme.*—This species is allied to the fallow-deer. The horns, the only parts hitherto discovered, shew that this animal, although nearly allied to the fallow-deer, must have been much larger. The horns occur in loose sand, and have been found in the valley of Somme in France, and also in Germany.

4. *Fossil Deer of Etampes.*—This species appears to be allied to the rein-deer, but much smaller, not exceeding the roe in size. The bones were found in abundance near Etampes in France, imbedded in sand.

5. *Fossil Roe of Orleans.*—This species was found in the vicinity of Orleans in France. It occurs in limestone, along with bones of the palæotherium. It is the only instance known of the remains of a living species having

been found along with those of extinct species. But Cuvier inquires, May not the bones belong to a species of roe, of which the distinctive characters lie in parts hitherto undiscovered?

6. *Fossil Roe of Somme.*—This species, the remains of which were found in the peat of Somme, appears to be very nearly allied to the roe.

7. *Fossil Red-Deer or Stag.*—This species resembles the red-deer or stag. Its horns are found abundantly in peat-bogs, or sand-pits, in England, France, Germany, and Italy.

He next describes the different species of fossil bovine animals in the following order.

1. *Aurochs.*—This species he considers as distinct from the common ox, and differs from the present existing varieties in being larger. Skulls and horns of this species have been found in alluvial soil in England, Scotland, France, Germany, and America.

2. *Common Ox.*—The fossil skulls of this species differ from those of the present existing races, in being larger and the direction of the horns being different. They occur in alluvial soil in many different parts of Europe, and are considered by Cuvier as belonging to the original race of the present domestic ox.

3. *Large Buffalo of Siberia.*—The fossil skull of this animal is of great size, and appears to belong to a species different from any of those at present known. It is not the common buffalo, nor can it be identified with the large buffalo of India, named *arnee.* Cuvier conjectures that it must have lived at the same time with the fossil elephant, and rhinoceros, in the frozen regions of Siberia.

4. *Fossil Ox, resembling the Musk Ox of America.*—The fossil remains of this species more nearly resemble the American musk ox than any other species, and have hitherto been found only in Siberia.

To these descriptions are added some observations on isolated bones of oxen found in alluvial soil in different parts of Europe. These bones are generally of the trunk or extremities. They are found along with the skulls and bones of the elephant, and appear to be nearly allied to those of the fossil buffalo of Siberia.

It would appear, from the facts just stated, that these fossil remains, both of deer and oxen, may be distinguished into two classes, the unknown and the known ruminants. In the first class our author places the Irish elk, the small deer of Etampes, the stag of Scania, and the great buffalo of Siberia; in the second class he places the common stag, the common roe-buck, the aurochs, the ox which seems to have been the original of the domestic ox, the buffalo with approximated horns, which appears to be analogous to the musk ox of Canada; and there remains a dubious species, the great deer of Somme, which much resembles the common fallow-deer.

From what has been ascertained in regard to the strata in which these remains have been found, it would appear that the known species are contained in newer beds than the unknown. Further, that the fossil remains of the known species are those of animals of the climate where they are now found: thus the stag, ox, aurochs, roe-deer, musk ox of Canada, now dwell, and have always dwelt, in cold countries; whereas the species which are regarded as unknown, appear to be analogous to those of warm countries: thus the great buffalo of Siberia can only be compared with the buffalo of India, the arnee. M. Cuvier concludes, that the facts hitherto collected seem to announce, at least as plainly as two imperfect documents can, that the two sorts of fossil ruminants belong to two orders of alluvial deposites, and consequently to two different geological epochas; that the one have been, and are now, daily becoming enveloped in alluvial matter; whereas the others have been the victims of the same revolution which destroyed the other species of the alluvial strata; such as mammoths, mastodons, and all the pachydermata, the genera of which now exist only in the torrid zone.

In the second Memoir we have a full, accurate, and

interesting account of the osseous breccia, or conglomerate, which occurs in the rock of Gibraltar, and in other limestone rocks and hills upon the coasts of the Mediterranean sea.

This breccia occurs in a grey-coloured compact distinctly stratified flœtz limestone, which abounds in the islands and on the coasts of the Mediterranean. It is not intermixed with the limestone, nor does it alternate with it in beds, but occurs filling up fissures, or in caves situated in it. It is composed of angular fragments of the limestone, of bones, usually of ruminating animals, generally broken, and never in skeletons, and land-shells, cemented together by a reddish-brown coloured ochry calcareous basis. The base is sometimes vesicular, and the vesicles are more or less completely filled with calcareous spar; and the spar sometimes traverses the conglomerate in the form of veins, or is more or less intermixed with it. Cuvier describes the osseous breccia of different tracts of country in the following order.

1. *Gibraltar.*—The mineralogical nature of this famous rock is well known, from the excellent description of it by our countryman Colonel Imrie. It is principally composed of limestone, and is frequently traversed by fissures, or hollowed into caves, in which the osseous breccia is contained. Cuvier found in it the bones of a ruminating animal allied to the antelope, and of a smaller animal of the order glires, which he conjectures may be-

long to the genus lagomys. All the shells contained in the breccia are fresh-water or land species.

2. *Cette.*—The breccia in this tract, like that of Gibraltar, occurs in limestone. In it Cuvier found bones of an animal not unlike the common rabbit; others of a species one-third less than the common rabbit; also bones of a species of mus, nearly allied to the field-mouse (mus arvalis, Lin.); of a bird of the order passeres; numerous vertebræ of a serpent somewhat resembling the coluber natrix; lastly, bones of a ruminating animal, probably of the same species as that found in the breccia of Gibraltar. Shells also occur. Three kinds are mentioned, viz. two helices, and one pupa, and all of them land-shells.

3. *Nice and Antibes.*—The limestone rocks of Nice contain this osseous breccia. Cuvier found in it bones of the horse, and of two species of ruminating animals. All the shells it contains are land species. The limestone rocks of Antibes, near Nice, also contain osseous breccia, in which Cuvier found remains of ruminating animals, apparently the same as those of Nice.

4. *Corsica.*—The limestone rocks containing the osseous breccia occur near Bastia, and agree in all their characters with that of Gibraltar. The osseous remains are principally of smaller quadrupeds, but they do not, like those of Cette, belong to species now living in the surrounding country; for Cuvier discovered there the head of an animal nearly resembling the *lagomys alpinus*, a

species which inhabits the wildest and most mountainous regions of Siberia, immediaetly under the snow line. He also found enormous quantities of the bones of a species of gnawer, somewhat resembling the mus terrestris of Linnæus, and of another very nearly allied to the water-rat.

5. *Dalmatia.*—The breccia is found throughout a great extent of limestone country. It agrees perfectly in its characters with that of Gibraltar. All the bones it contains, as far as Cuvier had an opportunity of examining, appear to be of the same size as those of the fallow-deer, and perhaps belong to the animal whose remains are found at Gibraltar. The remains of the horse have also been found in the breccia of this district; for the late John Hunter found the os hyoides of that animal in some masses of conglomerate from Dalmatia.

6. *Island of Cerigo.*—The only descriptions we have of this breccia, are those of Spallanzani and Fortis, from which it appears that it possesses the same characters as that of Gibraltar, &c. Spallanzani was of opinion that the bones belonged to the human species. Many years ago, Blumenbach refuted this opinion, and Cuvier shews that all of them belong to ruminating animals.

7 *Concud, near Teruel in Arragon.*—Bowles, in his Natural History of Spain, describes limestone rocks, containing an osseous breccia, as occurring at Concud. Cuvier is of opinion that it belongs to the same formation

as that of Gibraltar. It contains bones of the ox, ass, of a small kind of sheep, and many terrestrial and fresh-water shells.

8. *Osseous Incrustations in the Vicentine and Veronese.*—The natural history of these incrustations, or conglomerates, is still very imperfect. Cuvier found in them bones of the stag and ox.

Cuvier finishes his description of this osseous conglomerate, or breccia, with the following observations.

1. The osseous brecciæ have not been formed by either a tranquil sea, or by a sudden irruption of the sea. 2. They are even posterior to the last resting of the sea on our continent, since no traces are found in them of any sea-shells, and they are not covered by other beds. 3. The bones and the fragments of rock which they contain, fell into the rents of the rocks successively, and as they fell, became united together by the accumulation of the sparry matter. 4. Almost all the fragments contained in the fissures are portions of the bounding rock. 5. All the well ascertained bones belong to herbiferous animals. 6. The greater number belong to known animals, and to species that at present live in the neighbouring country. 7. The formation of these breccias, therefore, appears to be modern, in comparison of the flœtz rocks, and the alluvial strata, that contain remains of unknown land animals. 8. It is nevertheless still ancient, with respect to us, since nothing shews that such

breccia are formed at the present day; and some of them, as those of Corsica, contain also the remains of unknown animals. 9. The most essential character of this phenomenon consists more in the facility with which certain rocks have been split, than the matters contained in the fissures. 10. This phenomenon is very different from that exhibited by the caverns in Germany, which contain the bones of carnivorous animals only, spread over the bottom, in an earthy tuff, partly of an animal and partly of a mineral nature; although the rocks in which these caverns are situated do not appear to be very different from those which contain the osseous brecciæ.

The third Memoir treats of the fossil remains of the horse and hog.

Fossil remains of a species of horse are found along with those of the elephant, rhinoceros, hyæna, mastodon, and tiger. Cuvier confesses that he is not in possession of any means of ascertaining the species to which they belong. Very few remains of the hog tribe have been hitherto discovered in a fossil state; Cuvier mentions only a few bones and teeth of the common hog found in peat-mosses, or very new alluvial deposites.

In the fourth, fifth, and sixth Memoirs, we have a very interesting account of the fossil remains of the animals of the bear tribe. These are found in great abundance in limestone caves in Germany and Hungary. These caves vary much in magnitude and form, and are more or less deeply incrusted with calcareous sinter, which assumes a great variety of singular and often beautiful forms. The bones occur nearly in the same state in all these caves: detached, broken, but never rolled, and consequently have not been brought from a distance by the agency of water: they are somewhat lighter, and less compact than recent bones, but slightly decomposed, contain much gelatine, and are never mineralized. They are generally enveloped in an indurated earth, which contains animal matter; sometimes in a kind of alabaster or calcareous sinter, and by means of this mineral are sometimes attached to the walls of the caves. These bones are the same in all the caves hitherto examined; and it is worthy of remark, that they occur in an extent of upwards of 200 leagues.

Esper, who examined and described the caves of Gaylenreuth, on the frontiers of Bayreuth, informs us, that after passing through a succession of caves, he at length came to a narrow passage, which led into a small cave, eight feet high and wide, which is the passage into a grotto twenty-eight feet high, and about forty-three feet long and wide. Here the prodigious quantity of animal earth, the vast number of teeth, jaws, and other bones, and the heavy grouping of the stalactites, produced so dismal an appearance, as to lead Esper to speak of it as a fit temple for a god of the dead. Here hundreds of cart-loads of bony remains might be removed, bags might be filled with fossil teeth, and animal earth was found to reach to the utmost depth to which they dug. A piece of stalactite being here broken down, was found to contain pieces of bones within it.

Cuvier estimates, that rather more than three-fourths of these bones belong to species of bears now extinct; one half, or two-thirds, of the remaining fourth, belong to a species of hyæna, which occurs in a fossil state in other situations. A very small number of these remains belong to a species of the genus lion or tiger; and another to animals of the dog or wolf kinds; and, lastly, the smallest portion belongs to different species of smaller carnivorous animals, as the fox and pole-cat. We do not find in these caves any remains of the elephant, rhinoceros, horse, buffalo, or tapir, which occur so commonly in alluvial soil; and the palæotheria of the flœtz strata, the ruminating animals, and the gnawers, of the rock of Gibraltar, Dalmatia, and Cette, are never met with. Nor do we ever find the bears and tigers of these caves in alluvial soil, or in the fissures of rocks. The only one of the species found in these caves, and which is found elsewhere in other formations, is the hyæna, which occurs also in alluvial strata. It is quite evident that these bones could not have been introduced into these caves by the action of water, because the smallest processes, or inequalities, on their surface are preserved. Cuvier is therefore inclined to conjecture, that the animals to which they belonged must have lived and died peaceably on the spot where we now find them. This opinion is rendered highly probable from the nature of the earthy matter in which they are enveloped, and which, according to Laugier, contains an intermixture of animal matter, with phosphat of lime, and probably also phosphat of iron.

The following fossil species, found in these caves, are described by Cuvier.

1. Genus Ursus. Bear.

1. U. Spelaeus. The size of a horse, and is different from any of the existing species.
2. U. Arctoideus. Is a smaller species, and appears also to be extinct.

2. Genus Canis. Dog.

Of this genus several species are described as occurring in these caves: one species very closely resembles the Cape hyæna, and is about the size of a small brown bear; another species is allied to the dog or wolf; and a third species is almost identical with the common fox. *

3. Genus Felis. Cat.

One species of this tribe is described, and appears to be very nearly allied to the Iaguar.

4. Genus Viverra. Weasel.

Two species of this genus are described; the one is al-

* Blumenbach has lately described the remains of a fossil hyæna, nearly resembling the *canis crocuta*, which was found in marl along with remains of the lion and the elephant, between Osterode and Herzberg in Hanover.

lied to the common pole-cat, (v. putorius), and the other to the zorille, or pole-cat of the Cape of Good Hope.

In the eighth Memoir there is an account of several gnawers, principally of the genus castor, some of which are found in peat or alluvial soil, and others in slate.

The gnawers found in alluvial soil belong to the genus castor; and of these, two species are described: the one found in France appears very nearly allied to the common beaver, the castor fiber; the other, found on the shores of the sea of Azof, by M. Fischer, differs from the former, and is named *castor trogontherium.*

Two species of gnawers are described as occurring in slate: the one in the slaty limestone of Æningen, which Cuvier considers as belonging to the Guinea pig, cavia porcellus, or more likely to an unknown species of cavia or arvicola; the other is found in slaty rocks at Walsch, in Bohemia, in the circle of Saatz, and is nearly allied to the mus terrestris.

In the ninth Memoir there is an account of the osteology of the animals of the sloth tribe, inserted with the view of illustrating the nature of the fossil species. There are but two living species of the sloth genus, the Ai, or Bradypus tridactylus; and the Unau, or Bradypus didactylus. The fossil remains described by Cuvier are allied not only to these animals, but also to the ant-eater, or myrmecophaga, a genus allied to bradypus.

Two fossil species are described, viz. the Megalonix and Megatherium. 1. *Megalonix.* It is the size of an ox, and its bones were first discovered in limestone caves in Virginia in the year 1796. 2. *Megatherium.* This species is the size of the rhinoceros, and its fossil remains have hitherto been found only in South America. The first, and most complete skeleton, was sent from Buenos Ayres by the Marquis Loretto, in the year 1789. It was found in digging in alluvial soil, on the banks of the river Luxan, a league south east of the village of that name, about three leagues W. S. W of Buenos Ayres. *Plate 3d* gives a faithful representation of this remarkable skeleton, which is now preserved in the Royal Cabinet of Madrid. A second skeleton of the same animal was sent to Madrid from Lima, in the year 1795; and a third was found in Paraguay. Thus it appears, that the remains of this animal exist in the most distant parts of South America. It is very closely allied to the megalonix, and differs from it principally in size, being much larger. Cuvier is of opinion, that the two species, the megalonix and megatherium, may be placed together, as members of the same genus, and should be placed between the sloths and ant-eaters, but nearer to the former than to the latter. It is worthy of remark, that the remains of these animals have not been hitherto found in any other quarter of the globe besides America, the only country which affords sloths and ant-eaters.

The twelfth Memoir is very interesting. It contains the osteology of the Lamantin, and a discussion on the place which the Lamantin and Dugong ought to occupy in the natural method, and concludes with an account of the fossil bones of Lamantins and Seals.

The very general view we are now taking of this work, prevents us from doing more than noticing the fossil remains of these palmated quadrupeds, mentioned by our author. He describes two fossil species of *lamantin* found in the coarse marine limestone of the department of the Maine and Loire. This limestone, besides these osseous remains, contains abundance of sea-shells. In the same limestone, and in the same country, Cuvier found the fossil remains of a species of seal, nearly three times the size of the common seal, the phoca vitulina, and another somewhat less.

The thirteenth Memoir treats of the fossil remains of oviparous quadrupeds.

M. Cuvier, in this most valuable paper, describes the different species of living crocodiles, and dwells particularly on their osteology; and the fossil crocodiles are minutely examined and described. From the accounts here published, it appears that two extinct species of fossil crocodiles, nearly allied to the gavial, or gangetic crocodile, occur in a pyritical bluish-grey compact limestone, at the bottom of the cliffs of Honfleur and Havre; that one of these species at least is found in other parts of France, as at Alençon and elsewhere. 2. That the skeleton of a crocodile, discovered at the bottom of a cliff of pyritical slate, about half a mile from Whitby, by Captain William Chapman, probably belongs to one of these species. 3. That the fragments of heads of crocodiles found in the Vicentine, may be referred to the same species. 4. That fossil heads, found at Altorf, are different from those of the gavial, and have a longer snout than that of the animal of Honfleur, and may therefore belong to the other fossil species found in France. 5. That the remains of an unknown species of fossil crocodile was found near Newark, in Nottinghamshire, by Dr Stukely. 6. That the supposed crocodiles found along with fish in the copper slate, or bituminous marl slate, of Thuringia, are reptiles of the genus monitor. 7. Lastly, that all these fossil remains of oviparous quadrupeds belong to very old flœtz strata, far older than the flœtz rocks that contain unknown genera of quadrupeds, such as the palœotheriums and anopletheriums; which opinion, however, does not oppose the finding of the remains of crocodiles with those of these genera, as has been done in the gypsum quarries. *

In the fourteenth Memoir, Cuvier gives a particular description of the large fossil animal of the quarries of Maes-

---

* Sir Everard Hume has described, in the Transactions of the Royal Society of London for the year 1814, the fossil remains of an animal possessing characters partly of the crocodile, partly of the species of the class of fishes. It was found in a blue-coloured clayey limestone, named *Lias*, on the estate of Henry Host Henley, Esq., between Lyme and Charmouth, in Dorsetshire, and is now in the museum of Mr Bullock of London.

tricht. This species, which is one of the most celebrated of all the fossil species of oviparous quadrupeds, occurs in a soft limestone which contains flint, and the same kinds of petrifactions as are observed in the chalk near Paris. It is one of the most celebrated of all the fossil species of oviparous quadrupeds. Even so early as the year 1766 it had engaged the attention of inquirers, and up to the present day has not ceased to be an object of discussion and investigation among naturalists. Some have described it as a crocodile, others a whale; and it has been by some enumerated amongst fishes. Cuvier, after a careful study of its osteology, ascertained that this animal must have formed an intermediate genus between those animals of the lizard tribe, which have a long and forked tongue, which include the monitors and the common lizards, and those which have a short tongue, and the palate armed with teeth, which comprise the iguanas, marbres, and anolis. This genus, he thinks, would only have been allied to the crocodile by the general characters of the lizards. The length of the skeleton appears to have been nearly twenty-four feet. The head is a sixth of the whole length of the animal; a proportion approaching very near to that of the crocodile, but differing much from that of the monitor, the head of which animal forms hardly a twelfth part of the whole length. The tail must have been very strong, and its width at its extremity must have rendered it a most powerful oar, and have enabled the animal to have opposed the most agitated waters. From this circumstance, and from the other remains which accompany those of this animal, Cuvier is of opinion, that it must have been an inhabitant of the ocean. We have here then an instance of an animal far surpassing in its size any of the animals of those genera to which it approaches the nearest in its general characters; at the same time, that, from its accompanying organic remains, we find reason to believe it to have been an inhabitant of the ocean, whilst none of the existing lizard tribe are known to live in salt water. However remarkable these circumstances are, still they are not more wonderful than those we contemplate in many of the numerous discoveries in the natural history of the aneient world. We have already seen a tapir of the size of an elephant; the megalonix, an animal of the sloth tribe, as large as a rhinoceros; and here we have a monitor possessing the magnitude of a crocodile.

In the fifteenth Memoir there are descriptions of several species of fossil oviparous quadrupeds preserved in calcareous slate. This limestone, or calcareous slate, belongs to one of the newer floetz formations. The celebrated quarries in the valley of Altmuhl, near Aichstedt and Pappenheim, so rich in petrifactions, also the quarries of Aichstedt, are said to be situated in this limestone. The first petrifaction described by Cuvier is the celebrated "Homme Fossile" of Scheuchzer, which some naturalists, as Gesner, maintained to be the siluris glanis of Linnæus, but which is, in reality, nothing more than an unknown, and probably extinct, gigantic species of salamander or proteus. It was found imbedded in the slaty limestone of Æningen. The second petrifaction described was also affirmed by Scheuchzer to belong to the human species. The only bones found are vertebræ, and these Cuvier proves to be essentially different from those of the human species, and is of opinion that they probably belong to an animal of the crocodile tribe. They are found imbedded in limestone at Altorf. The third petrifaction noticed as occurring in slaty limestone, is an animal of the frog kind, found in the limestone of Æningen. Dr Karg, who has published a long account of the Æningen quarries, is of opinion, that this petrifaction is that of a common toad; whereas Cuvier is inclined to refer it to some species nearly allied to the bufo calamita. The fourth petrifaction is the celebrated flying reptile of Aichstedt, which some naturalists have taken for a bird, others for a bat, but which appears to form a new genus of the order Saurii of Dumeril, under the title of Ptero-dactylis. The only specimen of this petrifaction hitherto found, is that in the cabinet of the late Collini. Several years ago that cabinet was removed to Munich, and Cuvier applied to the celebrated journalist Von Moll, then at Munich, for information in regard to it. He informed him, it was no longer to be found; that is to say, it was concealed, to be out of the reach of French rapacity; for had Von Moll been so silly as to discover the hidden treasure, it might have found its way to Paris, to increase the stores of the National Museum. Collini was of opinion, that it belonged neither to the class of birds nor the tribe of bats, but might be a marine amphibious animal. Blumenbach supposes it to have been a web-footed bird; and the late celebrated naturalist Hermann arranged it along with animals of the class mammalia. Cuvier, with his usual consummate skill and sagacity, has rendered it highly probable that it is a species of a new genus of the order saurii.

The last Memoir of this volume, and which concludes this invaluable work, is "on the Fossil Bones of Tortoises."

1. Account of the fossil tortoises found in the environs of Bruxelles.

These remains are found imbedded in coarse marine limestone, at the village of Melsbroeck. They are different from any of the known animals of this division.

2. On the tortoises of the environs of Maestricht.

These are found in a coarse chalk, or limestone, in the mountain of Saint Peter. They are irregularly dispersed through the chalk along with many different marine productions and bones of the gigantic monitor, which have rendered this hill so celebrated in geology. All of these are remains of sea-tortoises, or chelonii, but of species different from any of those at present known.

3. On the tortoises of the slate of Glaris.

One species only is described, which appears to differ from any of those at present known, and is apparently a marine tortoise.

4. On the tortoises of the vicinity of Aix.

These fossil remains are of unknown species, and occur imbedded in a rock formation, apparently of the same nature as that around Paris.

Note K (B) § 28. p. 103.

Mineralogical Description of the Country around Paris.

As the very short description of the mineralogy of the country around Paris, in Note K (A), may not prove satisfactory to those who wish a more particular detail, we here insert a description, which, with assistance of the plate, will, we trust, enable the reader to form a distinct conception of all the important features of that remarkable district.

The country in the environs of Paris is entirely composed of newer flœtz rocks, of which the oldest, or lowest, is common chalk; the uppermost, or newest, alluvial. Interposed between these are nine different formations, principally of limestone, sandstone, and gypsum. The whole series of formations appear to be arranged in the following order, from below upwards.

1. The chalk formation, with flint.
2. Plastic clay, with sand (*argile plastique.*)
3. Coarse marine limestone (*calcaire grossier*), with its marine sandstone (*gres marine inferieur.*)
4. Siliceous limestone (*calcaire silicieux.*)
5. Gypsum and marl, containing bones of animals

(*marnes du gypse d'ossements*), with the lower, or first fresh water formation.

6. Marine marl, abounding in bivalve shells; and the upper layers, abounding in oyster shells.

7. Sandstone and sand, without shells.

8. Upper marine sandstone (*gres marine superieur.*)

9. Millstone, or buhrstone, without shells (*meuliere sans coquilles.*)

10. Upper or second fresh water formation, millstone, flint, and limestone (*terrein d'eau douce superieur, meuliere, silex, et calcaire.*)

11. Older and newer alluvial deposites (*Limon d'atterrissement.*)

FIRST FORMATION.

Chalk.

This chalk agrees, in external characters, with that found in other countries. It occurs in indistinct horizontal strata, in which we observe either interrupted layers or tuberose shaped masses of flint, which pass into the chalk at their line of junction, or kidneys of hard chalk, having the same shape and position with the flint. This formation is well characterised by the petrifactions it contains, which differ not only in the species, but sometimes also in the genus, from those that occur in the *coarse limestone.* Two species of belemnite occur in the chalk, and these appear to be different from those found in the limestone, and are considered to characterise it.

The chalk forms the bottom of the basin or gulph, in

which are deposited the different formations that occur around Paris. Its surface must have presented numerous inequalties before the present strata were deposited over it, because we observe promontories and islands of chalk rising through the newer formations.

SECOND FORMATION.

Plastic Clay.

All around Paris, we find the chalk covered with a deposite of plastic clay, which is dug and used in the manufacture of different kinds of pottery. This clay varies in colour, being white, grey, yellow, red, and black, sometimes contains a layer of sand, very rarely (only the purer varieties) organic remains, viz. cytherea, turritellæ, bituminous wood, and in some places fragments of chalk have been observed in it. It is neither intermixed with the chalk at its line of junction with it, nor is it more calcareous where in contact with that mineral, than at a distance from it; hence Cuvier conjectures, that it has been deposited after the chalk, and is therefore a separate formation.

THIRD FORMATION.

Coarse Marine Limestone, with its Marine Sandstone. *

This formation presents much greater variety than the chalk. Several different strata, or series of strata, such as limestone, clay-marl, limestone-marl, slate-clay, occur in it. These are arranged in a determinate order, and

---

* This coarse marine limestone is much used in Paris for building.

the strata of limestone are well characterised by the petrifactions they contain; the same system of strata always possessing the same species of petrifactions.

First System of Strata.

The lowest series of strata, or first system of strata, of the coarse limestone formation, is very sandy, and sometimes contains a substance resembling green earth; it is still better characterised by containing a great variety of well-preserved shells, many of which still retain the pearly lustre, and differ more from the present existing species, than those in the upper strata of this formation.

The following are the petrifactions enumerated by Cuvier and Brongniart, as occurring in it.

Nummulites lævigata, scabra, numismalis — These are always found in the lowest part of the bed.

Madrepora—At least three species.

Astræa—Three species at least.

Carophyllia—Three simple, and one branched species.

Fungites.

Cerithum giganteum.

Lucina lamellosa.

Cardium porulosum.

Voluta cithara.

Crassatella lamellosa.

Turritella multisulcata.

Ostrea flabellula.
Cymbula.

Second System of Strata.

The second system of strata is still very rich in shells; nearly all the bivalves found by M. Defrance at Grignon belong to it. It also contains rarely a few impressions of leaves and stems. The most characteristic petrifactions of this system of strata are the following.

Cardita avicularia.
Orbitolites plana.
Turritella imbricata.
Terebellum convolutum.
Calyptræa trochiformis.
Pectunculus pulvinatus.
Citheræa nitidula.
elegans.
Miliolites—It is very abundant.
Cerithium—Probably several species; but neither the lapidum and petricolum, nor cinctum and plicatum, which latter belong to the second marine formation which covers the gypsum.

Third System of Strata.

The third system of strata is already less abundant in petrifactions, and contains fewer species than the two preceding. The following have been observed.
Miliolites—Very rare.
Cardium Lima, et obliquum.

Lucina saxorum.
Ampullaria spirata.

Cerithium tuberculatum. mutabile. lapidum. petricolum.	Almost all the other species, with exception of the giganteum.

Corbula anatina?
striata.

Also impressions of the leaves of a fucus.

The strata of the second and third systems sometimes contain beds of sandstone, or masses of hornstone filled with marine shells. In some cases the sandstone takes the place of the limestone. Land shells and fresh water shells (*Limnæa et Cyclostomæ*) have been observed in this sandstone, but only where it lies immediately under the fresh water limestone. The sandstone and the hornstone, containing marine shells, rest either immediately on the marine limestone, or are contained in it. The following list contains the names of those species of petrifactions which occur most frequently in the sandstone.

Calyptræa trochiformis?
Oliva laumontiana.
Ancilla canalifera.
Voluta harpula.
Fusis bulbiformis.
Cerithium serratum.
tuberculosum.
Cerithium coronatum.
lapidum.
mutabile.

Ampullaria acuta, or spirati.
patula.
Nucula deltoidea.
Cardium lima.
Venericardia imbricata.
Cytherea nitidula.
elegans.
tellinaria.
Venus callosa?
Lucina circinaria.
saxorum.
Two species of oyster still undetermined; the one appears allied to *ostrea deltoidea*, the other to *ostrea cymbula*.

Fourth System of Strata.

This set of strata consists of hard calcareous marl, soft calcareous marl, clayey marl, and calcareous sand, which is sometimes agglutinated, and contains horizontal layers of hornstone, crystals of quartz, and rhomboidal crystals of calcareous spar, and small cubical crystals of fluor spar. Petrifactions occur very rarely.

FOURTH FORMATION.

Siliceous Limestone without Shells.

This formation occurs alongside the coarse marine limestone, on the same level with it, and in no instance either above or below it. It rests immediately on the plastic clay. It consists of strata, not only of a white limestone, but also of a grey, compact, or fine granular limestone, which is penetrated in all directions with silica; and its numerous cavities are lined with siliceous stalactites, or quartz crystals. A characteristic mark of this rock is its wanting petrifactions of every kind, both of fresh water and salt water. A species of millstone sometimes occurs in it, which appears to be the siliceous limestone deprived of its calcareous ingredient by some agent unknown to us. This millstone must not be confounded with that which occurs in the ninth formation.

FIFTH AND SIXTH FORMATIONS.

Gypsum of the first Fresh Water Formation, and the Marine Marl Formation.

This fresh water formation is not entirely of gypsum, but contains also beds of clay marl and calcareous marl. These are arranged in a determinate order when they all occur together, which is not always the case. They lie over the coarse marine limestone; and the gypsum, which is the principal mass of the formation, does not occur in wide extended plateaus, like the limestone, but in single conical or longish masses, which are sometimes of considerable extent, but always sharply bounded. Montmartre presents the best example of the whole members of the formation, and there three beds of gypsum are to be observed superimposed on each other.

The first bed consists of alternate layers of gypsum, solid calcareous marl, and of thin slaty argillaceous marl, or adhesive slate. The layers of gypsum are thin, and full of selenite crystals; and in the clay marl, or adhesive slate, occurs imbedded menilalite. Wherever this bed rests immediately on the sand of the marine sandstone containing shells, it contains sea shells. The former bottom of the sea, however, appears to have been frequently covered with a bed of white marl, on which the lower beds of gypsum rest, and this bed is filled with fresh water shells. The second bed resembles the first, and only differs from it in being thicker, and containing fewer beds of marl. The only petrifactions it contains are those of fishes. In the lower part of this bed we for the first time meet with single kidneys of celestine, or sulphat of strontian. The third, or upper bed, is by far the greatest, as it in several places is more than sixty feet thick. It contains few beds of marl; and in some places, as at Montmorency, it lies almost immediately under the soil. The lower strata of this upper gypsum contain flint, which appears to be intermixed with it, and to pass into it by imperceptible gradations—facts which shew their cotemporaneous formation. The middle strata of this bed split naturally into large prismatic concretions, with many sides. The uppermost strata, of which five generally occur, and extend to a great distance, are thinner than the others, and are intermixed with marl, and also alternate with beds of it.

Numerous quarries are situated in this upper gypsum, and which daily afford skeletons, or single bones of unknown birds and quadrupeds. To the north of Paris these are found in gypsum itself, where they are hard, and simply invested with marl; and to the south of Paris similar remains, but in a friable state, are met with in the marl which separates the beds of gypsum. Bones of tortoises, and skeletons of fish, are found in the same bed, and more rarely fresh water shells. This latter fact is highly important, as it shews the plausibility of the opinion of Lamanon, and other naturalists, who maintain, that the gypsum of Montmartre, and other hills in the basin of Paris, have been deposited from fresh water lakes. The occurrence of skeletons of quadrupeds particularly characterises the upper bed of gypsum, because remains of the same nature have not hitherto been discovered in the middle or lower beds of gypsum.

Beds of calcareous and clayey marl rest immediately over the gypsum. Woodstone, or petrified wood of a kind of palm tree, occurs in a white friable marl; and in quarries which are worked in it, remains of fishes and of shells, of the genera lymnæus, and planorbis, are met with. The two latter do not differ very much from those found in the marshes in France,—a fact which seems to shew, that this marl, as well as the subjacent gypsum, have been deposited from fresh water. In the numerous and thick beds of clayey and calcareous marl which rest over this white friable marl, petrifactions are so rare, that we cannot form any satisfactory opinion as to their formation.

The *first, or oldest fresh water formation*, of the vicinity of Paris, as already mentioned, contains gypsum, beds of marl that alternate with it and rest over it, and also the white friable marl, of which we shall now give a more particular account. It is in this marl that the fresh water shells, which principally characterise this formation, are found.

The first fresh water formation neither contains millstone nor flint, only menilite and woodstone. Over the beds of clayey and calcareous marl there rests a bed of yellowish slaty marl, three feet three inches thick. Kidneys of earthy celestine occur in the lower part of it; somewhat higher up we meet with a bed of small bivalve shells, which are referred to the genus Citherea, and between the uppermost layers of the marl other species of citherea, with cerites spirobites, and bones of fish, occur. This bed is not only remarkable on account of its great extent, (it has been traced ten leagues in one direction, and four leagues in another, and throughout its whole extent of the same thickness), but also because it marks the upper boundary of the first fresh water formation, and the beginning of a *new marine formation.* All the shells that occur in the marl above this bed belong to the ocean.

A great bed of greenish clayey marl, without petrifactions, rests immediately over the yellowish marl, and contains geodes, kidneys, and clayey calcareous marl, and also of celestine. Immediately over these follows a bed of yellow clay-marl, which abounds in fragments of marine bivalve shells, cerites, trochites, mactrites, cardites, venites, &c. and fragments of the tail of two species of ray have also been found in it.

The beds of marl which rest over these contain almost all the fossil marine shells, but only the bivalves; and in the uppermost bed of calcareous marl, immediately under the clayey sand, there occur two distinct beds of *oysters,* of which the undermost contains large and thick oysters, and the upper, which is sometimes separated from the under by a thin bed of white marl, without shells, numerous small, thin, and brown oyster shells. This latter bed of oysters is very thick, is divided into many layers, and is scarcely ever wanting in the hills of gypsum.

These oysters appear to have lived on the spot where we at present find them, because they are arranged as we find them in oyster banks in the ocean; and the greater number of them are whole, and with both valves. Lastly, M. Defrance found, near Roquencourt, at the height of the formation of the marine gypseous marl, rounded fragments of marly shell limestone, pierced with pholades, and with oyster shells attached to them. The formations sometimes terminate with a bed of clayey sand, in which no petrifactions occur.

Such are the beds which in general constitute the gypsum formation. We have described together the marine marl of the summit along with that of the gypsum, and the fresh water marl of the middle and bottom, because the beds nearly resemble each other, and constantly occur together.

In the following Table are enumerated the petrifactions that belong to the gypsum, and to the marine formation which rests on it.

Petrifactions of the Gypsum and the Marine Marl resting upon it.

FRESH WATER FORMATION.

Fossil quadrupeds in gypsum	Palæotherium magnum. medium. crassum. curtum. minus. Anoplotherium commune. secundarium. medium. minus. minimum. A pachidermatous animal, allied to the hog. Canis Parisiensis. Didelphis Parisiensis. Viverra Parisiensis.
Birds	Three or four species.
Reptiles . . .	Trionix Parisiensis, and another tortoise. A species of saurius, which appears to be a crocodile.
Fishes	Three or four species.
Molluscous animals	Cyclostoma mumia.
Upper white marl	Palms. Fragments of fishes. Limneus. Planorbes.

MARINE FORMATION.

Slaty yellow marl.	Cytherée bombée Spirobes. Bones of fishes. Cerithium plicatum. Cytherée planes. Bones of fish.	The shells of these petrifactions are generally in a powdery state, or we have only their mould or impression.

Green marl.	No fish.	
Yellow marl, mixed with brown slaty marl.	Parts of the ray. Ampullaria patula? Cerithium plicatum. cinctum. Cytherea elegans. semisulata. Cardium obliquum. Nacula margaritacea.	Almost all these shells are broken, and difficult to ascertain. The two species of cerites of the marine formation, which covers the gypsum, do not appear to occur any where else.
Calcareous marl, containing large oysters.	Ostrea hippopus. pseudochama longirostris. canalis.	The two beds of oysters are often separated from each other by marl without shells; and although we cannot say with any certainty whether or not the particular species here enumerated are shells that belong more to the one bed than to the other; yet it cannot be doubted, that the oysters of this marl do not occur in the coarse limestone, and that they are more nearly allied to the species at present living in our seas, than to those found in the limestone.
Calcareous marl, containing small oysters.	Ostrea cochlearia. cyathula. spatulata. linguatula. Ballanites. Shells of crabs.	

SEVENTH FORMATION.

Of Sandstone and Sand without Shells.

The sandstone with shells is one of the latest formations. It always rests on those already described, and in general is only covered with the millstone without shells, and the upper fresh water formation. * Its strata are often of considerable thickness, are intermixed with beds of sand of the same nature, and both are often so fine that they are used in manufactories.

EIGHTH FORMATION.

Upper Marine Sandstone and Sand.

This sandstone, or last marine formation, rests on the gypsum, marine marl, and even upon the sandstone and sand without shells. It varies in colour, compactness, and even in composition. Sometimes it is a pure sandstone, but friable, and of a red colour, as at Montmartre; sometimes it is a red-coloured clayey sandstone, as at Romainville; sometimes it is a greyish sandstone, as at Levignan; lastly, its place is occasionally occupied with a thin bed of calcareous sand filled with shells, which covers the great masses of grey, hard sandstone, and without shells, at Nanteiulle-Haudouin.

This sandstone contains marine shells, which are sometimes different from those found in the sandstone of the lower marine formation, and approach more to the spe-

* It appears, as we shall afterwards shew, that it is in some places covered by a formation of marine sandstone or limestone.

cies met with in the calcareous marl which surmounts the gypsum, as will appear from the following enumeration.

Shells found in the Upper Marine Sandstone.

Oliva mitriola.
Fusus? allied to longævus.
Cerithium cristatum.
lamellosum.
mutabile?
Solarium? Lam. Pl. viii. fig. 7.
Melania costellata?
Melania?
Pectunculus pulvinatus.
Crassatella compressa.
Donax retusa?
Citherea nitidula.
lævigata.
elegans?
Corbula rugosa.
Ostrea flabellula.

It thus appears, that there are three formations of sandstone in the vicinity of Paris, which are very nearly allied in their oryctognostic characters, but differ remarkably in their geognostic relations. The first, or lowest formation, is a member of the coarse limestone formation, and generally contains the same species of fossil shells as that rock. The second formation lies above the gypsum and the marine marl; it is the most extensive, and frequently appears at the surface. The third formation lies immediately over the second, and under the newest fresh water formation, and, like the first, contains a great number of bivalve fossil shells.

NINTH FORMATION.

Millstone without Shells.

This formation consists of iron-shot clayey sand, greenish, reddish, and whitish clay marl, and *millstone.* This millstone is a quartz, containing a multitude of irregular cavities, which are traversed by siliceous fibres, disposed somewhat like the reticular texture in bones. These cavities are sometimes lined or filled with red ochre, clay marl, or clayey sand, and they have no communication with each other. Most of the millstones found around Paris have a red or yellowish tint, but the rarer and most esteemed varieties have a bluish shade of colour. The bluish variety is the most highly prized, because it affords the whitest coloured flour; and a millstone of this kind, six feet and a half in diameter, sells at 1200 francs. We never observe in its cavities any siliceous stalactites, or crystallised quartz; and this character enables us to distinguish, in hand specimens, this millstone from that found in the siliceous limestone. It is sometimes compact. It has been analysed by Hecht in the Journal des Mines, No. xxii. p. 333, and appears to be almost entirely composed of silica. Another geognostic character of the millstone, properly so called, is the absence of all fossil animal and vegetable productions, whether of fresh or salt water origin.

It often rests on a bed of clay marl, which appears to belong to the gypsum formation; in some places it is separated from it by a bed varying in thickness, of sandstone or sand without shells. It is sometimes immediately covered with vegetable earth, but in other instances it has resting on it the upper fresh water formation, or the alluvial formation. *

TENTH FORMATION.

The Second Fresh Water Formation.

We have already described a formation which appears to have been deposited from fresh water, because the fossil animals it contains are analogous to those we find in

---

* The most extensive mass of this millstone occurs in the plateau which extends from La Ferte sous Jouarre (on the Marne, 16 leagues east from Paris) nearly to Montmirail; and here, near the first town, it has been quarried upwards of four hundred years for the excellent millstones it affords. The lower part of the plateau is marine limestone; the upper part, on the edges, and towards the Marne, of marl and gypsum; but in the middle, of an iron-shot and clayey sand, which forms a bed upwards of 60 feet thick. The millstone occurs in this great bed of sand, extends nearly throughout the whole plateau, and varies in thickness from three to five fathoms; but millstones cannot be made of every portion of the mass; hence we must not expect to find it throughout the whole bed. A bed of rolled masses of millstone, about a foot and half thick, lies over it; over this a thin bed of iron-shot sand, containing smaller pieces of millstone, and above this bed is one of sand, from 12 to 17 yards thick. If the stone rings when struck with a hammer, it will answer for large millstones. The bed never affords more than three millstones in the direction of its thickness. It frequently happens, that the fissures allow the workmen to extract the masses in a perpendicular direction, and these are the best. Millstones are formed by joining many of these parallepipedal pieces together, and confining the whole with an iron hoop. These pieces are exported from France, to England and America.

our fresh water lakes. This formation, which consists of gypsum and marl, is separated from another and more superficial fresh water formation, of which we are now to give an account, by a great marine formation.

This second fresh water formation, in the vicinity of Paris, consists of two sorts of stone, silex and limestone. These substances sometimes occur independent of each other; in other instances they are intimately mixed together. The nearly pure fresh water limestone is the most common; the next in frequency is the mixture of silex and limestone; the large masses of fresh water silex are the rarest. The silex is sometimes a nearly pure flint; sometimes approaches to pitchstone or to jasper: and, lastly, it has a corroded shape when it has all the characters of true millstone, but which is in general more compact than the millstone without shells. The limestone of this formation is white, or yellowish grey; sometimes nearly friable, like marl or chalk; sometimes compact and solid, with a fine grain and conchoidal fracture: the conchoidal varieties are rather hard, but easily broken into sharp-edged fragments, somewhat like flint, so that it cannot be cut. These characters apply only to the limestone near Paris; for, at a considerable distance, the limestone occurs very compact, of a greyish brown colour, which readily cuts and polishes. The limestone of Mont-Abusar, near Orleans, which contains bones of the Palæotherium, belongs to this formation. Even the hardest varieties of this limestone, after exposure to the air for a time, softens; and hence it is used as a marl for manuring the ground. All the varieties, both hard and soft, are traversed by empty vermicular cavities, whose walls are sometimes of a pale green colour. Where the siliceous minerals and the limestone are intermixed, the latter is always corroded, full of cavities, and its irregular cells are filled with calcareous marl. The essential character of this formation is, that it contains fresh water and land shells, nearly all of which belong to genera that now live in our morasses, but no marine shells ; at least in such places as are distant from the subjacent marine formation. The following is a list of those fossil organic remains that belong particularly to the upper fresh water formation.

Cyclostoma elegans antiquum.
Potamides Lamarkii.
Planorbis rotundatus.
    cornu.
    prevostinus.
Limneus corneus.
    fabulum.
    ventricosus.
    inflatus.
Bulimus pygmeus.
    terebra.
Pupa Defrancii.
Helix Lemani.
    Desmarestina.
Dicotyledonous wood, petrified with silica.
Stems of arundo or tipha.
Articulated stems, resembling the thorn.

Pediculated ovoidal grains.
Canaliculated cylindrical grains.
Olive-shaped bodies, with an irregular streaked surface.

The potamides, helicites, and limneus corneus, are the petrifactions that most frequently characterise this second fresh water formation, and the cyclostoma mumia has never been found in it. The first or lowest fresh water formation, on the contrary, has its characteristic petrifactions, the cyclostoma mumia, and Limneus longiscatus, and palludinus, but it never contains potamides, or helicites. It is remarkable that no bivalve shells occur in this formation, and that it contains numerous small roundish groved bodies, named Gyrogonites, which appear to be the fruit of a marsh plant of the Chara tribe.

This second fresh water formation extends for thirty leagues to the south of Paris, and has also been met with in the department of Cher, Alliere, Nievre, Cantal, Puy de Dome, Tarn, Lot, and Garonne, in the south-east of France, and more lately the same interesting formation has been discovered in the Roman states, in Tuscany, and in the vicinity of Ulm, Mayence, Silesia, in Estremadura, near Burgos, and other places in Spain.

ELEVENTH FORMATION.

Alluvial.

This appears also to be a deposite from fresh water. It consists of sand of many different colours, marl, clay, and even of mixtures of the whole three, which is intermixed, and coloured brown and black with carbonaceous matter, also of rolled masses of different kinds; and what particularly characterises it, remains of large organic bodies, large trunks of trees, and bones of elephants, oxen, deer, and other large mammalia. Although this formation is new, in comparison of those we have just described, yet it is of high antiquity in regard to man, as its formation extends to a period not far removed from the earliest periods of our history, when the earth supported vegetables and animals different from those that at present live in these or any other countries of the globe. The alluvial substances around Paris occur in two different situations, viz. *first*, in the present valleys; and, *secondly*, on the plains. In valleys they either cover the bottom, and then they consist of sand, loam, or peat; or they form in them wide extended plains, which lie high above the present river course, and then they consist of gravel and sand. It is difficult to distinguish the alluvial mud, situated at a distance from the valleys, from the fresh water formation, and it even, in some places, seems to pass into it. It appears, however, to be older than that of the valleys.

GENERAL OBSERVATIONS.

The eleven different formations now described are supposed to be partly of marine, partly of fresh water origin. Of these the three first above the chalk are of marine origin, and they cover the whole of the bottom of the basin of Paris. The gypsum, and accompanying marls, from the nature of the petrifactions contained in them, are conjectured to have been chiefly formed in fresh water. The next series of marls and sandstones, containing only marine shells, would seem to intimate that the sea had again covered the last deposited formations; and these, therefore, are of marine origin. Lastly, the upper fresh water formation shews this place to have been a second time converted into a lake, or covered with fresh water.

Several of these new flœtz formations, as already mentioned, have been discovered in other parts of Europe; and we may now add, that lately a series of rocks of the same general nature has been observed resting on the chalk formation in the south of England. The newer formations in this island were first pointed out, and described by Mr Webster, in a valuable memoir in the second volume of the Transactions of the Geological Society. That gentleman is of opinion, that two basins of chalk, filled with the newer formations, occur in the southern parts of England; one he names the Isle of Wight Basin, the other the London Basin.

1. *Isle of Wight Basin.*

The southern side of this basin extends from the highly inclined chalk at the Culver cliffs, at the east end of the Isle of Wight, to White Nose, in Dorsetshire, five miles west of Lulworth. The north side of it may be traced in that range of hills called the South Downs, extending from Beachy Head, in Sussex, to Dorchester, in Dorsetshire. The strata of which these hills are composed, dip generally from 15° to 5° to the south; the inclination varying in different places. The south side of the basin, therefore, must have been extremely steep, while the slope of the north side was very gentle. The closing of the basin at the west cannot be distinctly traced; but the east is now entirely open, the sea passing through it.

2. *London Basin.*

The south side of the basin is formed by a long line of chalk hills, including those of Kent, Surry, Hampshire, called the North Downs, extending through Basingstock to some distance beyond Highclere Hill, in Berkshire. Its western extremity is much contracted, and seems to lie somewhere in the vicinity of Hungerford. Its northwestern side is formed by the chalk hills of Wiltshire, Berkshire, Oxfordshire, Buckinghamshire, and Hertfordshire. The most southern part of this boundary has not yet been well determined. On the east it is open to the sea, the coasts of Essex, Suffolk, and Norfolk, being sections of the strata deposited in it. The dip of the chalk of the North Downs, from Dover to Guilford, is from 15° to 10°; but in the narrow ridge of chalk, called the Hog's Back, extending from Guilford to Farnham, the dip is very considerable, being about 45°. On the dip of the other sides, no observations have hitherto been made. The depth of the chalk below the surface at London must be very considerable; since, though wells have been sunk several hundred feet, it has never been reached; but at a few miles south of the metropolis, the chalk is frequently come to.

The formations described by Mr Webster as lying over the chalk, and in these basins in the south of England, are the following:

1. The lowest marine formation over the chalk, including the plastic clay, and sand, together with a particular clay, named the *London Clay*.
2. The lower fresh water formation, which rests immediately on the preceding formation.
3. The upper marine formation.
4. The upper fresh water formation.
5. Alluvium.

Chalk Formation.—The chalk which forms the sides and bottom of the basins, occurs distinctly stratified, and the strata vary in thickness from a few inches to several feet. The whole formation may be considered as composed of three great stratified beds, the undermost of which is named *chalk marl*; the second *hard chalk*, without flint; the third or uppermost, *soft chalk*, with flint. The *chalk marl* varies in colour, being grey, yellowish, and brown: it is softer than true chalk, and on exposure to the weather it rapidly disintegrates. It contains cotemporaneous nodules, and also beds of a more indurated marl, named *grey chalk*, from its dark colour. Like all argillaceous limestones, it possesses, in a considerable degree, the property of setting under water, when calcined and made into mortar. It contains the following petrifactions, viz. ammonites, scaphites, turrellites, trochites, and madreporites. The middle bed, the *hard chalk*, is in general harder than the bed above it, although Mr Webster remarks, that it appears from some observations he made in Dorsetshire, that the hardness does not always mark a particular bed, the flint chalk being in some places much harder than that without flints in others. It contains a greater variety of petrifactions than the chalk marl, as appears from the following list of the genera observed in it by Mr Webster. Several echini of the same families as those met with in the chalk with flint; but many of them, particularly the cassides, differing much in their forms from those found in that bed. Spines of echini; and particularly those described by Brard as resembling the belemnites. Patellites. Trochites. Serpulites, several species. Belemnites. Lima? Fish, too much mutilated to ascertain the genus. Palates, scales, vertebræ, and teeth of fish. Cancri.—The upper bed, the *soft chalk with flints*, forms the upper part of the formation, and is distinguished from the preceding by its softness, and always containing flints. It also differs from it in the petrifactions it contains, of which the following are enumerated by Mr Webster. Asteriæ. Echini of several families. Spines of the foregoing, resembling belemnites. Serpulites. Cardium. Spondylus. Ostrea, several species. Pecten, several species. Chama? Terebretula, many species. Alcyonia, sponges, and numerous unknown zoophytes. A ramose madrepore. Several species of minute encrini, figured by Mr Parkinson.

1. *Lower Marine Formation.*

This formation is separated into two great divisions, 1. Sand and plastic clay. 2. London clay.

1. *Sand and Plastic Clay.*—Of these two minerals the sand is the most extensive and continuous, and the clay occurs filling up basins and hollows in it. The clay varies in colour, being white, grey, yellowish-brown, and red. The white and grey varieties are potters clay. It sometimes contains beds of brown coal, from one foot to three feet thick; and beds of ironstone, and ferruginous sand, occur connected with it, and generally lying over it.

2. *London or Blue Clay.*—The bed which has received this name, is found immediately under the gravelly soil on which London is situated. Of all the strata over the chalk in the south of England, it is of the greatest extent and thickness; and the number, beauty, and variety of the petrifactions which it contains, render it the most interesting, and the most easily distinguishable. It consists generally of a blackish clay, sometimes very tough, at other places mixed with green earth and sand, or with calcareous matter. It contains also numerous flat spheroidal cotemporaneous nodules of hard marl, or clayey limestone, which lie in regular horizontal layers, at unequal distances, generally from four to forty feet apart. These nodules are well known by the name of Ludus Helmontii, or Septaria, from their being divided across by partitions or veins of calcareous spar, which are gene-

rally double. In their cavities are frequently found crystals of calcareous spar, and of heavy spar. The septaria are surrounded by crusts which contain a smaller proportion of carbonate of lime than the central part. They often contain organic remains.

Besides the clay, marl, sand, and carbonate of lime, of which the main body of this bed consists, several other substances are dispersed through it in smaller quantities. Of these the chief is iron pyrites, which is frequently the mineralising matter both of the vegetable and animal remains included in the blue clay. Silenite is also very abundant; and sulphat of iron sometimes effloresces, when the clay is exposed to the air, from the decomposition of the pyrites contained in it. Phosphat of iron is also sometimes found; and it abounds in Epsom salt, and in fossil organic remains.

In some places, as at Bognor, it assumes a new character; instead of a blue clay, we find a number of rocks now appearing as detached masses in the sea, though evidently forming portions of a stratum once continuous. The lowest part of these rocks is a dark grey limestone, or perhaps rather a sandstone, containing much calcareous matter, inclosing many organic remains belonging to the blue clay. The upper part is a siliceous sandstone.

This clay abounds in petrifactions, and of these the following copious list is given in Mr Webster's paper.

Organic Remains in the Lower Marine Formation above the Chalk in England.

NAMES GIVEN BY LAMARCK.	LINNÆAN NAMES.
Astroitæ.	Astroitæ.
Calyptrea trochiformis.	Trochus apertus. Brander.
Conus.	Conus.
Cyprea pediculus.	Cyprea pediculus.
Terebellum convolutum.	Bulla sopita. Brander.
Oliva.	Voluta.
Voluta spinosa.	Strombus spinosus.
musicalis.	luctator.
bicorona.	ambiguus.
crenulata.	Murex suspensus.
Buccinum undatum.	
Harpa.	
Cassis carinata.	Buccinum nodosum. Brand.
Rostellaria macroptera.	Strombus amplus.
Murex tripterus.	Murex tripterus.
tricarinatus.	asper.
tubifer.	pungens.
	contrarius.
	whirls the right way.
Fusus longævus.	longævus.
clavellatus.	deformis.
rugosus.	porrectus.
Pyrula nexilis.	nexilis.
Pleurotoma ?	
Cerithium gigantum.	Murex.
Cerithium, another variety, but too mutilated to ascertain the species.	Murex.

NAMES GIVEN BY LAMARCK.	LINNÆAN NAMES.
Trochus agglutinans.	Trochus umbilicaris. Brand.
monilifer.	nodulosus.
Solarium caniculatum, or Delphinula ?	Turbo, tab. 1. fig. 7. & 8. Brander. Turbo, tab. 1. fig. 7. Brand.
Turritella terebellatta	Turbo terebra.
imbricatoria.	editus.
multisulcata.	vagus.
Ampullaria patula.	Helix mutabilis.
Dentalium elephantinum.	Dentalium elephantium.
entalis.	entalis.
dentalis.	dentalis.
striatulum.	striatulum.
Serpula.	Serpula.
Nautilus imperialis.	
pompilius.	
centralis.	
Lenticulina rotulata.	
Nummulites lævigata.	
Pinna, 2 species.	Pinna.
Mytilus modiola.	Mytilus.
Pectunculus pulvinatus.	Arca glycemeris.
	noæ.
Cardium porulosum.	Cardium porulosum.
asperulum.	asperulum.
obliquum.	obliquum.
Crassatellata lamellosa.	Tellina sulcata.
Venericardia planicosta.	
Capsa rugosa.	Venus deflorata.

NAMES GIVEN BY LAMARCK.	LINNÆAN NAMES.
Chama lamellosa.	Chama squamosa.
calcarata	
sulcata.	
Ostrea edulis.	Ostrea edulis.
Pyrus bulbiformis.	
Caryophillia.	Turbinated madrepores.
Teredo navalis.	Teredo navalis.
Jaw of a crocodile.	
Testudo, or Turtle.	
Various Fish, but too mutilated to ascertain the species.	
Fish teeth, supposed by some to belong to the shark.	
Molar teeth of fish, called Bufonites.	
Palatum Scopuli, and other palates of fish.	
Tongue of a fish of the genus Raia.	
Tail of the Sting Ray.	
Scales of fish.	
Vertebræ of various species of fish.	
Cancer, above 20 species of crabs.	
——— Gammarus, or lobster.	
——— Crangon, or prawn.	

NAMES GIVEN BY LAMARCK.	LINNÆAN NAMES.
Wood, often pierced by the Teredo navalis, and filled with pyrites or calcareous spar.	
Fruits, branches, excrescences, ligneous seed vessels, and berries impregnated with pyrites.	

These fossil remains very nearly resemble those found in the lower marine formation in the basin of Paris,—a point of agreement of great importance, as it leads us to the probable inference, that the lower marine formation of the south of England belongs to the same deposite. This inference is strengthened, when we compare together the minerals of the different beds in the English and French formations.

Thus the plastic clay in the Paris basin agrees in most of its external characters with that found in the Isle of Wight and London basins; and further, both agree in the purer clays being destitute of organic remains, whilst the upper contains fossil cytherea and turritellæ. A species of coal also occurs in the lower strata of the Paris basin, and appears to be analogous to that found in a similar situation in the Isle of Wight basin; and the French sands agree in characters with those found in the Isle of Wight basin.

In the English basins there occur but few rocks that can be identified with the coarse marine limestone of the Paris basin. The rocks of Bognor appear to be the most easily referable to some of the beds of the coarse limestone of France; yet, in the Paris formation, there is no single rock possessing the same external characters as those exhibited by the London clay. But the London clay contains the same petrifactions as the coarse limestone; and if we could suppose a blending or mixture between the French plastic clay, which is blackish, and contains organic bodies, and the lower beds of the coarse limestone with its green earth and petrifactions, we should have a compound agreeing sufficiently near with the London clay under all its varieties; with this difference, that that of the French basin would have a greater proportion of calcareous, and ours of argillaceous matter. But with respect to the upper beds of the coarse limestone of France, no strata have as yet been discovered in England that correspond to them. *

2. *Lower Fresh Water Formation.*

It consists of a series of beds of sandy, calcareous, and argillaceous marls. Some of them appear to consist almost wholly of the fragments of fresh water shells, viz. lymneus, planorbis, cyclostoma, and others resembling helices, and mytuli. Hitherto, no alternation of these beds with marine strata has been observed. This forma-

* Webster's Geological Transactions, vol. ii. p. 209.

tion occurs in the Isle of Wight, but not in the London basin.

It is in this formation, in the Paris basin, that the gypsum beds are situated.

3. *Upper Marine Formation.*

Over the lower fresh water formation in the Isle of Wight, a stratum occurs, consisting of clay and marl, which contains a vast number of fossil shells wholly marine. Ten of the species agree with those found in the London clay, but they differ from them in their state of preservation. Most of them appear to have undergone but little change, and some of the species are even scarcely to be distinguished from recent shells.

Delicate marine shells, in a state of perfect preservation, occur in some parts; thus shewing that they could not have been brought from great distances, but must have lived near to the spots where they are now found. In other beds we meet with banks of large fossil oyster shells, the greater part of which are locked into each other in the way in which they usually live, and many have their valves united. It is therefore evident, that these oysters had not been removed from a distance to their present situation.

If we depend upon petrifactions as a means of enabling us to discriminate the different flœtz strata, we shall see reason to believe, that the last of the marine depositions in the south of England, are nearly allied to the upper marine formation in the basin of Paris.

In this bed in the Isle of Wight, Mr Webster found the following petrifactions.

NAMES GIVEN BY LAMARCK.	LINNÆAN NAMES.
Cerithium plicatum.	Murices.
lapidum.	Murices.
mutabile.	Murices.
semicoronatum.	Murices.
cinctum.	Murices.
turritellatum.	Murices.
tricarinatum.	Murices.
Cyclas deltoidea.	Venus.
Cytherea scutellaria.	Venus.
Ancilla buccinoides.	Voluta.
subulata.	Voluta.
Ampullaria spirata.	Helices.
depressa ?	Helices.
Murex reticulatus.	
Bivalve, apparently of the genus Erycina.	
Helicina ?	
Murex nodularius.	
Melania ?	
Natica Canrena.	
Ostrea, approaching to deltoidea.	
——— specific characters not evident, but different from the last.	

In the same formation at Harwich in Essex, the following petrifactions occur.

NAMES GIVEN BY LAMARCK.	LINNÆAN NAMES.
	Patella ungaria.
	lævis.
Patella spirorostris.	fusca.
Fissurella labiata.	
emarginula.	fissura.

NAMES GIVEN BY LAMARCK.	LINNÆAN NAMES.
Calyptrea sinensis.	Patella sinensis.
Eburna glabrata.	Buccinum glabratum.
	Murex corneus.
	erinaceus.
	contrarius.
	Trochus sulcatus.
	alligatus.
Ampullaria rugosa.	
Natica canrena.	
glaucina.	
Mactra.	
Venericardia senilis.	Arca senilis.
Lucina.	Venus gallina.
	Solen siliqua.
Pholas crispata.	
	Ostrea deformis.
Pecten plebeius.	
infirmatus.	
Balanus.	

Some of these, however, may belong to the lower marine clay.

Mr Webster appears to consider the Bagshot sand, which extends over a considerable tract of country in Surrey, and the blocks of granular quartz, named *grey weathers*, met with in Berkshire and Wiltshire, as members of this formation, and somewhat allied to the sand and sandstone of the upper marine formation in the Paris basin.

4. *Upper Fresh Water Formation.*

This formation also occurs in the Isle of Wight, in the hill of Headen, where it rests immediately on the last-mentioned, or upper marine formation. It is an extensive calcareous bed, fifty-five feet in thickness, every part of which contains fresh water shells in great abundance, without any admixture whatever of marine organic remains. The marl is soft, and easily affected by the weather, but includes a harder variety, which is so durable as to be employed as a building stone. Many of the shells found in this bed are quite entire, and these are intermixed with numerous fragments of the same species. They consist, like the lower fresh water formation, of several kinds of lymnei, helices, and planorbes; and from the perfect state of preservation in which they are found, appear to have lived in the places where they now are, the shells of these animals being so friable, that they could not have admitted of removal from their native situations without being broken.

Over this bed is another of clay, eleven feet in thickness, containing numerous fragments of a small non-descript bivalve shell. Upon this lies another bed of yellow clay without shells, and then a bed of friable calcareous sandstone, also without shells. To this sandstone succeed other calcareous strata, containing a few fresh water shells. In these are parts of extreme compactness, and other parts contain masses of a loose chalky matter, most of which are of a round form; and among these also are many beds of a calcareous matter, extremely dense, and much resembling those incrustations that have been formed by deposition from water on the walls of ancient buildings in Italy. Through all these last strata are veins, frequently several inches in thickness, of radiated calcareous spar. It contains the following fossil shells:

Planorbis, much resembling that which Brongniart says approaches to P. cornu.
Planorbis, two other species.
Planorbis, much resembling P. prevostinus.
Ampullaria.
Cyclostoma.
Limneus longiscatus.
acuminatus.
corneus.
Gyrogonites is the petrified seed of a species of chara.

This formation is the latest of the flœtz rocks hitherto observed in this island, and it agrees nearly with its corresponding formation in the Paris basin, with this difference, however, that it contains no siliceous beds.

5. *Alluvial Formations.*

The fresh and salt water flœtz rocks already described, are in many places covered with various alluvial deposites. In the Isle of Wight and London basins, the alluvium, besides the vegetable earth, clays, marls, and sands, contains a vast quantity of rounded quartose pebbles, of various kinds and sizes, which are irregularly distributed, in some places forming thick beds, mixed with clay, sand, and small fragments of flints; in others are mixed with shells of various kinds, and sometimes almost without any other substance. This compound is named *Flint Gravel.* *

* Some of these pebbles are evidently fragments of the flint originally belonging to the chalk formation; but other varieties are of calcedony and hornstone. Another remarkable class of siliceous pebbles is found either mixed with the flints, calcedonies, and hornstones, or alone, or cemented into a pudding-stone. These, according to Mr Webster, appear to have been originally formed of concentric coats, or layers of different colours, which vary in almost every specimen. The colours are for the most part yellow, brown, red, bluish, black, grey, and white; but these run into each other by an infinite number of shades. Others are spotted, or clouded with different tints, and have much the appearance of Egyptian pebbles. They take an excellent polish, and are then often extremely beautiful. These last appear rather more to resemble agates than chalk flints. They are never found of large size, seldom exceeding two inches in diameter, and generally are not more than one inch. They are of an oval or flattened form, which appears to have been their original figure, although they have evidently been subjected to a certain degree of attrition. The well known pudding-stone of Hertfordshire is composed of these concentric pebbles, imbedded in a basis of granular quartz. These concentric pebbles, like the imbedded masses of flint in chalk, of agate in trap, and of felspar in porphyry, are to be viewed as having been formed at the same time with the rock in which they were formerly included.

The alluvial deposites in the south of England also contain fossil bones of quadrupeds; and these, according to Mr Webster, are of different dates. The most ancient are entirely petrified, and where found in gravel, are conjectured to have been washed out of the strata in which they were originally imbedded. Of this kind are probably remains of the mastodon, mentioned by Mr Parkinson. The next class contains the bones of the elephant, rhinoceros, hippopotamus, and the Irish elk, which are frequently accompanied with marl, and fresh water shells. They are, however, not petrified; and though generally in a state of decay, yet are sometimes quite perfect. They are particularly abundant in Suffolk and Norfolk, but have also been found at Brentford, in the Isle of Sheppey, and several other places. Other bones of ruminating animals as those of the horse, ox, and stag, not different from the living species, are frequently dug up at small depths, and are covered by peat, gravel, loam, &c. Similar organic remains occur in the alluvial strata, over the new flœtz-rocks around Paris.

The discoveries of Cuvier, Brongniart, and Webster, of which we have now given a pretty full account, have added a most interesting and curious set of rocks to the geognostic system. They have shewn, in a satisfactory manner, the existence of the partial or local formations first pointed out by Werner, and afterwards by Von Buch. They have connected, more nearly than heretofore, the alluvial with the flœtz formations, and have thus rendered more complete the series of rocks which extends from granite to gravel. Not the least interesting of the consequences resulting from the discoveries of these Naturalists, is the extension they give to our views in regard to the former nature of the animal world, and of the changes it has experienced during the different periods of the earth's formation.

THE END.

Printed in the United States
By Bookmasters